Frontispiece. — PANORAMA IN THE CASCADE NATIONAL FOREST, OREGON.

All of the natural resources, soil, water, minerals, and forests should be put to their best permanent use. The broad conservation movement in the United States had its beginning with Forestry.

ELEMENTS OF FORESTRY

FREDRICK FRANKLIN MOON
NELSON COURTLANDT BROWN

GREENWORLD
PUBLISHERS

First Indian Reprint 2001

ISBN 81-87421-23-1

Published by
Greenworld Publishers
8/217, Indira Nagar
Lucknow - 226 016 (UP)
Tel. : 701433

Exclusively Distributed by
International Book Distributing Company
Chaman Studio Building, 2nd Floor,
Charbagh, Lucknow - 226 004 (India)
Tel. : Off. : 450004, 450007, 459058 Res. : 711172, 342829
Fax : 0522-458629
E-Mail : ibdco@sancharnet.in

Printed at :
Army Printing Press
33, Nehru Road, Sadar Cantt
Lucknow - 226 002
Tel. : 0522- 481164, 480546
E-mail : pressneversleeps@rediffmail.com

TO

Theodore Roosevelt

WHO AS PRESIDENT OF THE UNITED STATES AROUSED
THE AMERICAN PEOPLE TO THE IMPORTANCE
OF FORESTRY AS A NATIONAL ISSUE AND
HELPED TO MAKE POSSIBLE ITS
RAPID PROGRESS IN THIS
COUNTRY

PREFACE.

The development of the American forest policy within the last two decades has been no less than remarkable. When it is realized that the first forest reserves were set aside barely twenty years ago and that the present gross area of the National Forests is 186,000,000 acres, some conception of the material strides may be obtained.

The most striking change, however, has taken place in the improvement in public sentiment. Twenty years ago, scarcely one man in a hundred knew what forestry meant, while it now constitutes an important part of the nation's news.

The associations and organizations in the United States now concerned with forestry comprise the following:

31 states with forestry departments;
17 states having conservation or kindred commissions;
2 national conservation organizations;
1 national forestry association;
23 state and local forestry organizations;
42 associations for the protection of timber land, etc.

In addition, there are 23 institutions giving degrees in forestry, 11 schools with forestry courses of at least one year's duration, and 42 schools giving short courses in forestry.

Realizing the educational awakening that is taking place along forestry lines, it was felt that an up-to-date textbook was needed, broad in its scope and containing general information on all phases of the subject.

In the state agricultural schools and colleges, forestry

should be taught to every student of agriculture. The land problem will soon be pressing and, in some regions, the ultimate forest area is very large in comparison with the arable soil. With the rapid increase in the tilled areas, the farm woodlots will become of greater importance, and scientific handling should then be assured.

Appreciating the fact that forestry will be complementary to agriculture, an especial effort has been made to suit these conditions. The bibliography will be of particular advantage to teachers without a forestry degree, and with its help any one with a scientific education should be able to give an effective synoptical course.

The forest regions are covered in a suggestive fashion only, and no doubt many details can be supplied by the individual instructors from the references given. No chapter has been included on dendrology, as it is too large a subject to be touched upon in a general textbook. There is plenty of material dealing with the identification and silvical qualities of the forest trees, published privately or by the U. S. Forest Service, available in practically any college library.

It has been the chief object of the authors to gather data from sources not readily available and to present them in a form easily grasped by the average student. Much of the material and data have been compiled from other sources, Federal and State bulletins, etc., but the presentation is entirely original.

Most of the illustrations were obtained from the Forest Service; to its members and to the friends who have aided with criticisms and suggestions, the Authors extend grateful acknowledgment.

FREDERICK FRANKLIN MOON.
NELSON COURTLANDT BROWN.

SYRACUSE. NEW YORK,

CONTENTS.

PART I.

PART II.

PAGE

ELEMENTS OF FORESTRY

PART I.

CHAPTER I.

INTRODUCTION.

Forestry Defined and Explained.

Definition. — Forestry is the art of raising repeated crops of timber on soils unsuited to agriculture, and properly utilizing the products of the forest. Under certain circumstances, however, agricultural land may be used to grow trees to good advantage, as in raising catalpa in the Middle West.

It will be seen from this definition that forestry and agriculture are not antagonistic, but that each is complementary to the other and a proper appreciation and use of both is necessary to the proper use of land in any country.

Forestry Means Use. — On account of the way in which the forest policy in the United States was built up, the idea that forestry means locking up the forestry resources without using them has gained wide credence. This is an unfortunate misunderstanding, for forestry means freest and fullest use, consistent with the maintenance of a permanent timber supply.

It is an unfortunate circumstance that funds for the development and even protection of the National Forests were so tardy in forthcoming that many citizens got the idea that the "Forest Reserves," as they were then called, were set aside for the purpose of preventing extensive exploitation.

This misconception has aroused much antagonism to the policy of the Forest Service.

The opening of the agricultural portions of the National Forests to settlement, the wide use of the grazing areas within their boundaries, and with timber sales increasing year after year, the public is coming to appreciate the extent and character of this use and as a result sentiment favorable to the Forest Service policy is rapidly taking the place of the old antipathy and distrust.

The National Forests are now supplying free fuel, fencing, etc., to the value of $191,000 to 38,000 people living in and around the boundaries; timber amounting to over $1,500,000 in value was sold during the current year; future sales to the extent of $4,500,000 were consummated; 1,500,000 people used the National Forests as a camping and recreation ground; over 10,000,000 head of stock grazed upon their grassy parks and 1175 cities and towns and 324 irrigation and power projects drew their water from streams having their head-waters within the National Forests. Surely this is free and proper use.

Forests may serve other purposes than the production of timber. For instance, bark for tanning or naval stores may be the end in view; or the indirect influences of the forest in retarding run-off or checking the force of drying wind (shelter-belts) are extremely important. However, the growing of wood is the chief end of the forest, for without timber, railroads, cities and in fact civilization would have been impossible. The other ends served by the forest are incidental for the most part.

Forestry Versus Lumbering. — Forestry has been considered by many to be synonymous with lumbering. This is not the case, however, but lumbering is, in reality, an important part of forestry.

The chief point of difference between the lumberman and the forester is that the latter recognizes growth as a function of the forest while the lumberman entirely overlooks this fact and makes no provision for the future.

The forester regards a piece of ground too steep or stony to be used for agriculture as true forest soil, which should produce successive crops of timber perpetually. If this area is now supporting a stand of timber, so much the better since growing timber is forest capital that can be harvested as soon as the timber crop is ripe and a new growth of trees started immediately.

The lumberman, on the other hand, is more apt to consider a tract of virgin timber in the same light that a mine owner regards his mineral lode. It contains much valuable material that must be marketed with all despatch and then the land can be thrown aside as worthless.

The difference between lumbering and forestry has been summed up in these words "Lumbering means devastate the country and move on" while forestry means "Develop the non-agricultural acres to their maximum and hold on."

It is gratifying to see that in many parts of the United States lumbermen are realizing that timber cannot be classified as an inexhaustible resource, and they are now taking steps to cut and utilize it as conservatively as market conditions will warrant.

From the above definitions it can be readily seen that forestry is not based on sentiment nor upon the desire to preserve the forests for their beauty, but that the entire forestry movement is founded on sound economic principles. It endeavors to make soils which are unfit for anything else produce a commodity that is practically indispensable. In addition, these non-agricultural areas which would otherwise lie idle are thus made to yield a revenue that will pay taxes and interest on the capital which they represent.

In summing up, forestry and agriculture are coördinate and lumbering is a phase of forestry, for the mature timber must be cut so that a new crop may be started.

FIG. 1. — HOMESTEADERS CLAIM INSIDE IDAHO NATIONAL FOREST, IDAHO.

The Forest Service desires each part of the National Forest to be put to its proper use. The land in the foreground is valuable for agriculture, hence it has been thrown open to settlement.

Forestry Versus Arboriculture. — Arboriculture and forestry are also often confused and by many are used interchangeably. There is a wide difference, however, between the two lines of activity.

Arboriculture means the raising of trees singly or in groups for any purpose whatever, while forestry is concerned with growing trees in large bodies, called forests, for the production of timber.

The wide use of the term City Forester has added to the confusion. It really should be City Arboriculturist, as the raising of single trees for shade purposes is his aim and not the production of timber, but nevertheless usage has almost sanctioned the former term.

Need of Forestry in the United States.

In spite of the large areas which at the present time are supporting wood growth, a definite clear cut forest policy is badly needed in the United States for the following reasons.

First. We are cutting our timber about three times as fast as it is growing.

Second. Our per capita consumption is unnecessarily high, being 260 cubic feet against 40 cubic feet in Germany and 12 cubic feet in Great Britain.

Third. Our per acre production has sunk so low under the poor forest management until it is only about one-fourth of the possible yield.

Fourth. Already the end is in sight for some species of timber and the virgin supply of forest material in the United States will be practically exhausted by the year 1950. This, of course, means that from then on we will be compelled to use second growth timber which has been grown within the memory of man. Consequently, it behooves us now to take stringent measures to prevent waste; to protect our forests against fire and to increase the growth of our forests to the maximum; to get the new crop of trees started so that when the virgin forests have been exhausted it will be possible for the next generation to get forest products sufficient to supply their needs.

The national government as well as the states are taking steps to forestall this situation by setting aside large areas of non-agricultural land to be used for producing timber for future needs.

Development of Forestry.

General. — Forestry, or rather arboriculture, was practiced by the ancients with some degree of success; in fact, the ancient Greeks and Romans were much more skilled in the handling of woodlands than is generally known. Plantations in some instances were made close to the cities for local supply, and the selection system and coppice system of forest management were roughly practiced. Forestry was considered at that time as a part of agriculture and the naturalist of that period dwelt upon the management of forests at considerable length in their agricultural treatises. Pliny goes into the technique of silviculture and among other methods mentions planting, grafting, layering and pollarding.

Their science of forest management in some respects was decidedly advanced but, on the other hand, considerable superstition was intermixed with actual forest knowledge.

Growth in Germany as Typifying Development.—Of all the modern nations Germany has most thoroughly mastered forestry practice in all its details, and it is to the sturdy Teuton that we owe a heavy debt for the development of our national forest policy. Some of the early educators and administrators were of German birth and education and many American foresters have received a portion of their training in Germany.

The study of the development of forestry in Germany is of more than ordinary interest because in reaching its present state of development it has passed through the various evolutionary stages which every country must experience. While

German methods cannot be blindly copied in the United States owing to different economic and climatic conditions, nevertheless many of the fundamental principles can be appropriated and their mistakes avoided.

The chief use of land in the early days was for hunting and

FIG. 2. — MARKING TIMBER FOR SALE, COCONINO NATIONAL FOREST, ARIZONA.

The United States owns 600 billion board feet of timber or $\frac{1}{5}$ of the national supply. It is the duty of the government to cut out the mature so that young growth may be started and future generations be supplied with this indispensable commodity.

for pasture. Timber was plentiful, apparently inexhaustible, hence the forest was cut with no thought of the future. This custom of squandering natural resources is common to mankind in general and was the rule in Germany in the tenth century as well as in the United States in the eighteenth and nineteenth centuries.

During the twelfth and thirteenth centuries, however, owing to the fact that local wood supplies were running short, and stringencies were in evidence, certain restrictions were made concerning the management of the forest land. Fuel was limited to the actual needs of the tribe. The timber to be cut was marked by some one in authority and bark peeling and burning for potash were forbidden. It is interesting to note the similarity of these restrictions to those put in force in the State of Maine from the years 1640 to 1665.

This era passed to that of forest extension. During the fourteenth century considerable artificial reforestation was practiced. Nuremberg in 1368 planted between 200 and 300 acres of spruce and fir and in 1491 the city of Seligenstadt agreed to plant 20 to 30 acres with oak each year. During the fifteenth century a timber shortage was feared and consequently more drastic measures to preserve the forest supply of timber were put in force. Pasture in newly cut areas was forbidden and a diameter limit was set by the city of Brunswick in 1483. Thus we see that the diameter limit is among the earliest of restrictive methods to be put in force in any country. It appears logical but in practice may do considerable damage in even aged stands.

Toward the end of the fifteenth century, forest fire laws were enacted but not until the beginning of the eighteenth century did a general forest policy have its beginning.

Under men like Cotta, Hartig and Heyer, the present elaborate system of forest management which distinguishes the German forest practice was built up, and to the selection and coppice systems used by the ancients the shelterwood and clear cutting systems were added.

State Regulation. — In this connection it might be interesting to trace the growth of state regulation, the endeavor of those in authority to regulate the cutting of privately owned timber. The first measure of this sort was passed in Bavaria in 1516. Brunswick toward the latter part of the sixteenth century and Württemburg in the early part of the sixteenth century also passed measures regulating the control of timber cutting on private lands. Private properties were either directly placed under technical administration or permission to cut the timber must be secured from those in authority. Therefore the recent effort of the part of legislation in Maine and Louisiana to regulate the cutting of timber on privately owned land for the benefit of the state at large, had its counterpart in Germany and Bavaria as early as the sixteenth century.

Forestry Abroad.

Germany. — At the present time the total forest of Germany contains about 35,000,000 acres; a little over three-fifths of an acre for each inhabitant. Intensive forestry is the rule and the result of their painstaking efforts during the 150 years that forest management has been in force are shown in the ever increasing yield and profits.

France. — The state forest property in France amounted to 2,900,000 acres, which is less than 12 per cent of the total forest area. The Departments and Communes own about 3,400,000 acres.

The most notable achievements of the French foresters have been:

1. The reclamation of waste lands and fixation of sand dunes. 275,000 acres of drifting land have been reclaimed during the last century and in the region called the Landes, composed of shifting sands and marshes, the French forest officers have reclaimed over 1,750,000 acres. Draining first and then planting was the method practiced in the moist

locations and a total sum of $10,000,000 was expended in this district. It is estimated that the forests resulting from this investment are now worth upward of $100,000,000.

2. Reforestation of mountain slopes has been carried on on a gigantic scale and at enormous expense. During the French Revolution large areas of the steep slopes were denuded and streams which had previously been easily con-

FIG. 3.—CATTLE GRAZING WITHIN THE DESCHUTES NATIONAL FOREST, OREGON.

One sixth of the total meat supply of the United States graze for a part of the year within the National Forests.

trolled, became absolutely unmanageable. Altogether 1462 streams became liable to flood and 1,000,000 acres of mountain land were exposed to erosion in addition to vast areas subject to overflow farther down the valley. The work was started in 1860 and already an area of over 500,000 acres has been acquired and about half of it has been planted at a cost of $13,000,000. So far, 163 torrents have been cured and 645 greatly improved. The budget for the improvement of these mountain streams calls for an expenditure of $600,000 a year until 1945. It is estimated that the repair of these streams will cost upward of $50,000,000.

Switzerland. — The Swiss republic while not possessing the forest area of either France or Germany has brought commercial forestry to a high state of development. Out of the total area 2,000,000 acres of forest, no less than 1,300,000 are owned by the cities and towns against 90,000 acres owned by the cantons. The state and town forests are managed intensively so that the yield of the former often goes a long way toward paying the expenses of the city government.

The most notable example of this type of ownership is the Sihlwald, which has been owned by the city of Zurich since 853, has been cut according to a specific plan since 1384 and now ranks as the best managed forest in all Europe. In spite of the large amount of money that is spent in tending the forest crop, the financial returns are remarkably high, no less than $7.25 per acre per year being the return from these non-agricultural acres, lying in the steep valley of the Sihl.

The use of forests as aids in flood prevention is highly appreciated and the Swiss government has already spent over $1,000,000 in subsidizing repair operation to control the mountain torrents by means of construction of stone and concrete barrages in addition to reforestation of the drainage areas.

In land of such steep slope where every acre of level agricultural soil is indispensable the working out of a combined system of reservoir construction and reforestation has kept every part of the land bearing to its utmost capacity and has resulted in an economic saving that is incalculable.

Forestry in the United States.

The forestry movement of the United States is of comparatively recent growth compared with the centuries of technical forest management in Germany, France and Switzerland, but in reality there were some attempts at forest protection, etc., in colonial times.

Early History. — William Penn as early as 1682 inserted a clause in the titles he gave, to the effect that one acre should be kept in forest for every five acres cleared. In view of this early attempt it seems like the irony of fate that Pennsylvania should have a larger per cent of unseated and denuded lands than any other eastern state.

Federal legislation looking toward the preservation of a timber supply first took the form of an attempt to provide naval timber in 1799 and 1817. This was followed by another act in 1831 which further protected species valuable for naval construction by providing a punishment for their destruction. For about one-half century this statute was the only one in any way providing protection for the national resources.

These efforts at restrictive legislation were chiefly negative in character and had but little bearing on the forest policy of to-day. The real beginning of technical forestry in the United States can be traced to a paper read before the American Academy for the Advancement of Science in 1873 by Dr. F. B. Hough. A memorial was presented to Congress the next year by this Association and Dr. Hough was appointed Forest Agent in the Department of Agriculture in 1876. This agency which was chiefly a clearing house for information became a Division (1886) and later a Bureau of Forestry in the Department of Agriculture (1897).

The funds for the maintenance of this division were extremely meager and public opinion was decidedly apathetic. Despite this lack of financial and moral support splendid investigative and propagandist work was done in these early days and much of the later development of the Forest Service was planned and started in the dark days of the late 80's.

Extension of Forest Reserve Policy. — In 1891 a law was passed empowering the President to set aside from the national domain, forest lands whether wholly or in part covered with timber as public reservations.

President Harrison set aside the first reserves and altogether there were withdrawn from the national domain during his administration over 13,000,000 acres. President Cleveland followed the lead of his predecessor and withdrew 22,000,000 acres of the best timberland that remained unappropriated. Unfortunately the best that was then owned by the Federal Government could not compare in quality and location with the superb stands of timber that had been acquired by the far-seeing lumbermen of the Rocky Mountain and Pacific Coast States long before.

As a result of this tardy recognition on the part of the National Government of the importance of acquiring a Federal forest domain, the forest area now owned by the government in most cases lies far back in the mountains, remote from railroads and markets, while private concerns own the splendid timber in the valleys and on lower slopes.

This act of President Cleveland aroused a storm of protest from the western lumber operators and land owners in general. They had seen the public domain loosely handled and so easily acquired for such a long time that they did not look favorably upon the withdrawal of this enormous resource, rich in possibilities. In fact, much of the opposition to the National Forest policy has been engendered and accelerated by these disgruntled lumbermen and stock men, who at that time lost the opportunity to make millions out of the property that belongs to all the people.

Under McKinley's administration about 7,000,000 acres were added, bringing the total up a little over 46,000,000 by 1901.

When Colonel Roosevelt took the presidential chair, the withdrawals increased enormously. Heartily in sympathy with anything concerning the public welfare and federal ownership, President Roosevelt not only added to the forest area owned by the nation but with forestry as a lever set in motion the conservation movement which has since so marvellously extended its activities. In both he was greatly aided by the eminent chief forester, Mr. Gifford Pinchot.

When Colonel Roosevelt left office the area within the National Forests was 194,500,000 acres and the bulk of the forest land in the Coast and Rocky Mountain States still owned by the people was under the control of an efficient, non-political, technically trained body of men, the Forest Service, and was being administered by them for the best interest of the present and future generations. To Roosevelt and Pinchot the future American citizen will owe a heavy debt.

Under President Taft the forestry movement extended itself but little as far as area was concerned but internal progress was quite rapid. The technical work went on; trails and roads were built; and telephone lines constructed for the better protection and management of the forest. Classification of the forest land has been practiced for some time which has resulted in the elimination of land better suited to agriculture or grazing than to timber raising.

Present Situation. — At the present time the total area within the National Forest boundaries is 186,616,648 acres, situated in the United States, Alaska and Porto Rico, divided into 165 national forests, containing with the State Forests about one-fourth of the total forested area, one-fifth of the national timber supply and having a cash value of upward of two billions of dollars. This superb timbered area is now most efficiently administered by the Forest Service and every part is open to the freest possible use. Mineral claims may be patented within the forest boundaries; land more valuable for agriculture than for timber production can be acquired under act of June 11, 1906. Open grass lands are extensively grazed, supplying feed to one-sixth the total meat supply of the United States.

Finally ripe timber is sold as fast as purchasers can be found, under such conditions that the future supply of forest products is assured. All these phases of activity are being carried on actively, efficiently and honestly. In fact, the Forest Service furnishes a striking example of efficient government management conducted upon an enormous scale.

Comparison with Conditions Abroad. — While in view of the preceding figures it would appear that enormous strides have been made in working out a forest policy in the United States within the last ten years, in reality we are still far behind most of the continental countries. The reasons for this tardy development are many. In the first place our forest policy does not date back more than forty years at the most while Germany for instance has been working and evolving systems of forest management for the last 150 to 200 years. Then, too, economic conditions are not comparable. The labor costs in America are extremely high and the value of stumpage is comparatively low. No less an authority than Mr. Pinchot has stated that with few exceptions no timber has been sold in the United States at a price equal to the cost of producing it. However, unless we do receive a higher price for stumpage and manufactured products, we cannot spare any of the meager profits to reinvest in our forest land to produce larger and better crops of timber. Finally while extensive lumbering is no doubt largely controlled by economic conditions nevertheless custom, often blindly followed by the lumberman, is responsible for much of the waste and poor management in American forests.

The amount of money and labor expended on the average German forest is startling to an American; there is on the average one employee to about 300 acres of forest. These figures give a marked contrast to the situation as it appears in the United States where owing to the short sighted economy of Congress only about 2½ cents per acre per year is spent in protecting and improving the National Forest domain and only enough funds are appropriated to permit one ranger or guard to every 125,000 acres. In Germany it is quite common for the man in charge to spend anywhere from $1 to $4 per acre per year taking care of the forest property and as a result of his careful administration a net revenue of from $3 to $10 per acre per year is obtained.

State Work. — At the present time 13 states have a central body concerned entirely with the administration of a forest policy within their boundaries, while in 15 additional states the forestry activity is either entrusted to a commission or board which is also concerned with agriculture, fish and game or all of the natural resources of the state. In 3 states there is a single forest commissioner and forester, making a total of 31 states in which definite strides are being made toward the development of a clear-cut forest policy, and of the number, 20 have technically trained state foresters in charge of the work. The annual appropriations for forestry work range from several hundreds of dollars to over $300,000.

Fourteen states have acquired lands to be managed as state forests, the areas ranging in size from 2000 to over

1,800,000 acres. The total controlled by the states amounts to 3,400,000 acres.*

In addition to practicing forestry on the publicly owned lands, most states are developing a splendid system of coöperative work with private owners. Advice is given concerning the management of farm woodlots, etc., seedlings are raised and distributed to citizens at a low cost, forest experiment stations are conducted by the states either in connection with the forests or at educational institutions. This propagandist work is telling not only in concrete forms, as evidenced by the fact that over 1,000,000 acres of private land are now planted to forest trees, but in the less apparent but enormously important form of greater and increasing interest in forestry and conservation. Granges, civic organizations and women's clubs, especially throughout the east, are interested in forestry and conservation, and since under any free government a movement cannot advance faster than public opinion will sanction, it seems that the cumulative effects of the widespread educational policy are now appearing and that the future bodes well for forestry in the United States.

BIBLIOGRAPHY.

CLEVELAND, TREADWELL. What Forestry has Done. Forest Service Circ. 140.

CLEVELAND, TREADWELL. Status of Forestry in the United States. U. S. Forest Service Circ. 167.

FERNOW, B. E. History of Forestry.

PINCHOT, GIFFORD. Primer of Forestry, Parts 1 and 2. Published as Farmers' Bulletins 173 and 358. U. S. Dept. of Agric.

VAN HISE, CHAS. R. Conservation of Natural Resources.

ZON, RAPHAEL. Forest Resources of the World. U. S. Forest Service Bulletin 83.

ZON, RAPHAEL. Future Use of Land. U. S. Forest Service Circ. 159.

* For a complete tabular statement of status of State, municipal and private Forestry, see Appendix.

FIG. 4.—A CONIFEROUS SEEDLING SHOWING ROOTS, STEM, CROWN AND FRUIT.

The strong tap root is in marked contrast to the slender laterals.

CHAPTER II.

THE TREE.

Definition.

A tree is a plant with a single woody trunk that does not branch for some distance above the ground.

Even if 20 or more feet in height a plant that branches directly at the ground is considered a shrub, although some plants that are shrub-like in the northern part of their range attain tree form in the southern part, owing to a longer growing season, greater precipitation, etc.

Parts and Functions.

A tree is composed of roots, trunk or stem, branches, leaves, flowers, fruit and seed.

Roots. — The roots are of two kinds, surface and tap, depending on their shape and depth of penetration. The surface roots, as their name implies, run along under the ground; the tap roots strike almost directly down and in deep soils may attain great length.

Of the tap-rooted species walnut, hickory and oak are the most prominent eastern species; longleaf and bull pine are the most notable of the southern and western species. Spruce, birch, elm, western larch, lodgepole pine and hemlock are surface-rooted species. Other trees, like cottonwood and red gum, may develop a combination system with either the surface or tap root more developed, depending on the nature of the site.

The root habit of a species has an important bearing on the system of forest management which can be used. For instance, a heavy thinning in eastern spruce or lodgepole pine is attended with danger, since, owing to their shallow root system, windfall is likely to follow.

FIG. 5.— UPTURNED WESTERN YELLOW PINE, MONTEZUMA NATIONAL FOREST, COLORADO.

In addition to possessing a stout tap root, this specimen had a strong set of lateral roots. The roots in the picture have a spread of over 20 feet in addition to the broken portion remaining in the ground.

The roots serve to anchor the tree and at the same time supply it with food. The nourishing parts of the roots, the root hairs, are located just back of the growing point and the older and larger parts merely anchor the tree and transmit the nutrient solutions, the "raw sap," taken in by the root hairs. The soluble salts in the soil are taken in through the cell walls of the root hairs and are carried up through the vessels and tracheids of the sapwood. The reason for the rise of the sap is as yet not clearly understood, being variously ascribed to osmotic forces, atmospheric pressure, the action of the living cell and the suction of transpiration, with none of these hypotheses clearly proven.

Root growth is greater in poor than in good soil and root extension is greater in a dry than in a wet season for the reason that the roots, under adverse circumstances, must seek further to provide the proper amount of food.

Stem. — The stem, like the root, is composed of inner and outer bark, heart-wood and sapwood. Increase in diameter is achieved by means of the cambium layer which lays on a ring each year. Occasionally when the growth of the tree has been checked in mid-summer, owing to drought or defoliation, an extra ring may be laid on. The false rings are fainter, and more or less irregular and are generally easy to detect.

The heartwood is located in the center of the trunk and is more or less inert material whose chief function is to strengthen and stiffen the trunk. It does not transmit any of the sap either up or down the tree, which may be proved by examining some of the hollow trees so common in our pastures. Such trees live for years with the heartwood completely rotted away but if the cambium and sapwood were interrupted by a deep cut as in "girdling" the course of the elaborated and raw sap would be interrupted and death would almost immediately ensue. In certain species there is a limited amount of interchange between the heartwood and sapwood but for the most part it is the sapwood alone which is alive and which transmits the nutrient solutions from the soil to the leaves.

The function of the bark is to distribute elaborated food and for protection. Like the trunk it adds a ring each year. The rings are much flattened by pressure, and are hard to detect.

Leaves. — The leaves are the stomachs of the tree to which the thin watery solutions are carried. Here the water is combined with the carbon dioxide taken in through the stomata (those minute openings occurring chiefly on the under side of the leaves) to form sugars and starches. The carbon dioxide is split up and combined with water, the extra oxygen and water are given off and the process of assimilation takes place in the presence of chlorophyll.

As stated previously the flow of the sap is not clearly understood, but it is known that the raw sap is conducted to the leaves in the sapwood, that it is there elaborated, and that it then passes down through the cambium and bast to the growing portions of the tree.

The part of the watery solution that is not combined passes out of the leaves in the form of a vapor, leaving the salts behind in the leaves and twigs. This process of evaporation, called transpiration, accounts for the fact that the leaves and twigs contain more mineral salts than any other part of the tree.

Consequently when the humus, which is composed of partially decomposed leaves and twigs, is not destroyed by fire or burned up by direct exposure to the sun's rays, the fertility of the forest site remains unimpaired. In fact the soil becomes richer year after year, because the roots are continuously bringing up mineral salts from the subsoil, which are left behind in the leaves in the process of transpiration, and when the leaf fall occurs this fertility is restored to the surface soil, adding to its mineral content, as well as vastly improving the physical quality by the presence of humus. In addition the bulk of the tree is composed of carbon dioxide and water, therefore a very slight demand is made on the soil in growing a crop of timber. This explains why land which is far too poor to cultivate profitably can raise a crop of timber satisfactorily. In a general way trees only require one-fourth to one-half as much fertility as the average field crop.

In addition to the transpiration process, which results in the giving off of oxygen and water vapor, the tree breathes like any other living organism; that is, it takes in oxygen and emits carbon dioxide. This breathing of the tree goes on continuously but is not nearly so active as the assimilative process. Consequently much more carbon dioxide is used than is emitted and much more oxygen is given off than is consumed.

Considerable moisture is used for the building up of the plant and large quantities of water are needed for good growth. Consequently in regions where rainfall is scanty, trees may be present but their form will be decidedly poor.

An English investigator states that a stand of beech 150 years old will consume 400,000 gallons of water per acre per year. This seems an extremely large amount but in terms of rainfall is not excessive; in addition it must be remembered that beech is a moisture-loving species.

Tree Characteristics.

Of the various characteristics the following are the most important and will be considered in turn:

Occurrence.	Tolerance.
Local extension.	Duration of life.
Form.	Quality of wood.
Soil and moisture requirements.	Reproduction.
Growth.	Resistance.

Occurrence. — Continental distribution of a species depends primarily on heat. Different species need various degrees of heat for their proper development and the carrying on of the different life processes, and there is a certain average temperature for each process in a given species.

For instance, the amount of heat required to germinate the seed of a certain species is much higher than for another, and on the ability of a plant to reproduce depends its range and natural extension.

Moisture also has an important bearing and on large continents while heat may affect the range north and south, moisture is apt to limit tree growth east and west; the interior portions of the continent may have a rainfall that is too scanty to support tree growth.

The altitudinal range of trees is really a question of heat since height zones due to the average temperature during the critical season can be clearly marked. The northern limits of tree growth varies from 55 degrees north latitude on the Labrador coast to 70 degrees north on the coast of Alaska; the presence of the Japan current accounting for this difference.

The timber line on high mountains grows lower as we pass from the tropics toward the pole, since the zone of the same annual average temperature descends lower as colder climates are approached.

In the Himalayas the timber line is found at 14,000 feet above sea level; in the Alps at about 7500 feet; and in the Adirondacks at about 4000 feet above the sea. For the same reason a species occurring at a given elevation in the southern Rockies will be found at lower elevations as the northern part of its range is approached.

Ecologists are at variance regarding the length of the growing season necessary for tree growth; some claim that 50° F.

must be the average temperature during the four growing months, while others insist that it is the temperature during the six critical weeks that count.

Concerning the southern limit for any tree it is generally believed that excessive heat is rarely, if ever, fatal, but a northern tree transplanted to a southern climate is apt to produce infertile seed, which limits the natural extension of the species. Also on account of the absence of the normal resting period, continuous growth may weaken the tree so that it will succumb to some destructive agency.

To a certain degree the continental distribution is a matter of chance, either present or past. For instance fewer tree species are found on the European than on the North American continent, owing to the fact that following the glacial epoch trees in Europe were unable to regain the ground they had lost because they were unable to extend themselves again beyond the lofty mountain ranges running east and west.

On the North American continent the mountain ranges run north and south and the trees while temporarily pushed out of their ordinary habitat did not have high mountain barriers to hold them back after the recession of the ice sheet. As a result we have over 500 tree species in America against 100 in Europe.

Local Extension. — By local extension is meant the range of trees within their geographical distribution.

Northern trees toward the southern part of their range are apt to be found on the northern and northeastern slopes; southern trees are apt to be found on slopes of a southerly aspect toward the northern part of their range.

This point is of importance in the management of certain species as it may be useless to attempt the natural regeneration of a tree on a given site when it is not congenial, if considered in connection with its local distribution in that region.

Form. — Each species of tree has a form that is quite typical but only a tree which has ample growing space, a pasture elm for instance, will assume this form.

When a tree is growing in the dense forest the form will be affected by the crowding of its neighbors and as a consequence the typical form is altered.

Aside from the amount of growing space the age of a tree has an important bearing on its form. Most trees attain the greater part of their height growth before there is any large diameter growth, consequently up to the age of 35 to 40 the average tree is apt to be tall and slender.

Then after the period of most rapid height growth is past, the tree begins to put on flesh, as it were, and diameter and volume growth commence to increase. So marked is this characteristic that it is possible to roughly estimate the age of a tree by the amount of taper it possesses, for if a tree carries its diameter well up into the crown it is safe to assume that it is fairly mature.

The quality of site on which the tree stands also affects its form. If the soil is thin with little moisture, the tree is apt to be short and scrubby, while if the tree is rooted in deep rich soil with plenty of moisture available, tall full boled trees are the rule. In fact the forester often gauges the quality of his forest site by the form and, especially, height of the trees growing on it.

Light soils produce as a rule long straight boles of average diameter and medium crowns while heavy soils commonly produce trees with heavy round crowns and thick boles.

Soil and Moisture Requirements. — Different species vary widely in the demands they make on fertility and water content of the forest soil.

In fact one of the first things a forester should realize is that an extremely close watch must be kept on the condition of the forest soil in order to encourage the growth of the desired species.

German foresters claim that the forest soil is just what the man in charge makes it but this remark cannot be taken too literally, since a gravelly soil with excessive drainage can never support a good thrifty stand of a moisture-loving species.

In each portion of a forest, the soil should be studied so that the proper species may be encouraged on the site where it can grow thriftily and make the kind of timber that the market demands.

Hardwood species are inclined to be much more exacting than conifers since they need about four to six times the moisture and about double the fertility that the less exacting evergreens require. Of the coniferous species spruce requires much fresher and richer soil than pine, for example, and among the pines there are species which are less exacting than others. The red or Norway pine can grow on dry gravelly soils where white pine would perish and it is believed that the Scotch pine of Europe surpasses even the red pine in its drought-resisting qualities.

Such species as walnut, maple, beech, etc., need a deep, rich, moist soil for best growth and on a site of this description they can compete successfully with softwoods so that the latter will eventually be crowded out. Such trees as poplar, gray birch, black oak, and in the West the pinon and juniper, are decidedly less exacting in their soil and moisture requirements.

In each case there is a balance between the need for food and for sunlight. If there is plenty of food available sunlight may be reduced to a minimum and *vice versa* if a tree is found thriving under shade in a poor location it may be considered extremely tolerant.

Trees are easily satisfied in their youth and become more exacting as they grow older. There are many instances of plantations started on sandy soil which thrived up to the age of forty and then with no especial reason their growth rate began to decrease and the stand soon fell prey to fungus and insect attacks. It is because the site no longer sufficed and their resistance was lowered.

Growth. — By growth we mean the increase of a tree in height, diameter or volume. Of these height growth is of first importance in the early life of a tree, since on the early height of a seedling depends its fate in the struggle for existence.

Height growth varies widely according to species, locality and treatment. Primarily it depends on the energy of the leading shoot; the terminal bud is larger and the energy therefore is greater. Any influence which favors the formation of a hardy bud and later helps this bud to develop can be counted on to favor height growth. In many cases reproduction will get started under the shade of the older trees and for a few years appear to be thrifty. Unless the canopy is opened the seedlings will lose their thrifty appearance, the foliage will change to a pale yellowish green and the buds instead of being large and vigorous will become small and weak. When this is the case light must be given immediately or the reproduction will die.

Of all the factors influencing height growth, light perhaps is of paramount importance since it is absolutely necessary to assimilate plant food. Direct light is not absolutely necessary at first and by some it is believed to retard height growth, for the lengthening of the shoot takes place in the night or in the early morning hours. The energy of the sun's rays supplied, either by direct or diffused light, must be available, else growth of any kind will cease.

Heat also is indispensable for growth. Below 0° C. growth practically ceases although certain species of Arctic plants grow through the snow. Above 40° to 50° C. growth also stops on account of excessive heat. The optimum temperature is considered to be from 20° to 30° C., depending on the individual habits of the species.

Water is absolutely essential for life and growth since cell division is dependent on the presence of water within the cells; in fact 95 per cent of the growing tissue of a plant is water. Vigorous height growth is an index of sufficient water as well as plant food in the soil, and lack of moisture nearly always causes scrubby growth.

Aside from the effect of the site on the height growth of a given tree, the species itself is extremely important. Of all the trees found in the United States, the exotic eucalyptus is capable of attaining the greatest height, but of the native species the sequoia has that distinction, reaching a maximum height of from 320 to 350 feet.

A tree increases in height most rapidly during the early part of its life, and, after the bulk of the height growth has been attained, the diameter and volume growth tend to increase. Intolerant trees are generally the fastest growing; sprouts are also faster growing than seedlings of the same species owing to the large root system already formed, but the growth of sprouts soon culminates.

Extreme density is apt to decrease height growth, owing to the reduction of light and excessive competition between the roots, but reasonable crowding is beneficial since with unlimited growing space the bulk of the growing energy of the tree is dissipated in branch production rather than in producing a large, straight, cylindrical bole. One investigator claims that the rapid height growth of trees in close stand is due, not to the limiting of branch production, but to the fact that electrical currents are set up when the lower branches interlace.

A tree increases in diameter by laying on a ring of wood each year deposited by the cambium layer and the increase in volume depends on height and diameter growth together. On the whole it might be said that they are subject to the same laws as height growth, but species is relatively of more importance. The increase in volume is apt to begin earlier with intolerant trees since the struggle for existence is terminated earlier with light-demanding individuals, leaving fewer trees standing on the site to utilize the solar energy and plant food. Trees having the same height may differ in diameter as 1:4, depending on the closeness of the stand.

Since between 50 to 60 per cent of green wood is water it is found that the dry weight of wood produced by trees of different species is much closer than the volume produced, in the same time. Consequently trees of light specific gravity are the best volume producers and since tolerant trees can stand closer together and can assimilate food from the air and soil equally well, it follows that the largest volume per acre is produced by tolerant trees of light specific gravity.

The size of the crown as a rule is ordinarily a good index of the amount of volume growth, for a tree with a large crown can elaborate more food to be distributed about the trunk. One theory concerning the distribution of growth is that woody material is laid on in the greatest amounts where it is needed most; that stress in the tree acts as an irritant as it were, and that extra amounts of wood are laid on where the strain and irritation is the greatest. Trees that are standing in the open have thick trunks at the base where the strain is the greatest and trees leaning downhill have the widest rings on the downhill side.

After Pinchot, trees may be classified according to their size as follows:

Up to 3 feet in height — Seedlings (provided they are of seedling origin.)

3 to 10 feet in height . . . . Small saplings.
10 feet in height, 4 inches in diameter. . . Large saplings.
4 to 8 inches in diameter . . . . Small poles.
8 to 12 inches in diameter . . . . Large poles
1 to 2 feet in diameter . . . . Standards.
Over 2 feet in diameter . . . . Veterans.

Tolerance. — The ability of a tree to endure shade is called tolerance. The amount of shade that different trees can bear varies widely according to the species.

Of all the factors that influence tolerance the amount of food and moisture available is the most important, for if a plant has enough food and just enough light to assimilate it, the tree can thrive, but if the amount of food in the soil is limited the tree needs all the solar energy possible to elaborate a sufficient amount of food.

In this connection the length of the growing season is of great importance because a long season permits the tree to lay up a large amount of food in the parenchyma on the days that diffused light is available; therefore we often find that trees in the southern part of their range are much more tolerant than they are in the northern part. The red cedar commonly regarded as an intolerant tree in the North grows quite frequently in the South under the shade of the yellow pine.

Trees possess inherent characteristics concerning the amount of shade they can bear. Species like the pitch pine, aspen, western yellow pine, etc., are ranked as intolerants and require full sunlight for their development. Trees like white pine, yellow birch and chestnut are in the middle of the scale; while beech, hemlock, hard maple, western red cedar, etc., are markedly tolerant and can grow in quite heavy shade. Some of the important trees might be classified as follows regarding tolerance.

HARDWOODS

Tolerant	Intermediate	Intolerant
Hard maple	Chestnut	Black locust
Beech	Red oak	Tulip poplar
Red maple	Butternut	Gray birch
Black gum	Ash	Black cherry
White elm	Black walnut	Hickory

CONIFERS

Tolerant	Intermediate	Intolerant
Balsam fir	Loblolly pine	Larches
Hemlock, eastern and western	White pine	Lodgepole pine
Spruces, eastern and western	Scrub pine	Western yellow pine
White fir	Red fir	Red cedar
	Douglas fir	Pitch pine
	Western white pine	Norway pine

Intolerance is a characteristic which tends to increase with age. White pine seedlings can grow and thrive for a few years in the shade of older trees. However, the increasing need of light soon becomes apparent. Their leaves become pale yellow instead of a deep lustrous green, and their buds become smaller and less vigorous in appearance.

Shade-bearing species usually have the following characteristics:

The forest has a dense canopy, the trees have thin leaves, thin bark, thick sapwood; branches are persistent, and on the ground a thick layer of humus is usually found, owing to the density of the canopy which shades the soil and retards normal disintegration of humus by largely excluding the sun's rays.

Tolerant trees, as well as those having a low requirement for plant food and moisture have a great advantage in seizing and holding forest sites. Owing to their ability to stand shade they will ultimately crowd out more intolerant species and occupy large areas, provided a proper dissemination of seed is possible. Light seeded trees, like the aspen and gray birch of the Northeast, often seed in over large areas after a forest fire, but they merely constitute the "nurse crop" and are generally crowded out by the slower growing but more tolerant individuals like the white pine and spruce.

Duration of Life.— The length of time which a tree lives depends largely on a combination of internal and external conditions. Some scientists claim that a tree never dies of old age, but that its death is always due to external factors. However interesting that statement may sound it is true, nevertheless, that there are certain definite ages which given species do not seem to exceed, and ordinarily there is a certain average age which may be put down for each species.

Some species are remarkably long lived. The sequoia, attaining an age of 3500 years, is the longest lived American tree. Short-lived species are also noticeable in the way they tend to disintegrate at certain ages. For instance a stand of aspen generally breaks up at about 90 years of age and in fact the best yields are obtained when the stand is cut at about 25 years of age, as the growth rate falls off early.

The bulk of a tree lasts but a few years and then becomes inert; the sapwood which is alive gradually turns into heartwood, which serves chiefly to stiffen the trunk of the tree. The growing points of a tree, however, are alive and continually expanding.

To attain this maximum age, normal for a given species,

conditions must be favorable. There should be sufficient room for proper crown development and there should be also sufficient space for the roots to develop without excessive competition. If the site is sufficient and if there is enough light, the life of a tree will be greatly prolonged. In fact trees growing in the open are considerably longer lived than those of the same species grown in forests.

Of short-lived trees poplar, balsam fir and gray birch are the best examples of the eastern species which reach their climax between 60 and 80 years. Lodgepole pine for the Rocky Mountain species is considered short-lived, maturing at 120 to 200 years. Beech and maple may be considered mature at an age ranging from 300 to 400 years and white oak, Douglas fir, sugar pine and western larch average about 500 years. The sequoias, as before mentioned, are examples of long lived trees, as the big tree reaches an age estimated from 3500 to 5000 years and the redwood may live from 1000 to 1500 years.

Quality of Wood. — The quality of wood laid on by any species is very largely a matter of the individual. For instance, spruce wood is composed of long fibers which makes a light elastic wood extremely valuable both for wood pulp and musical instruments. Oak produces a dense heavy wood which can withstand considerable crushing force, and consequently the two are used for entirely different purposes, the difference in structure affecting the use.

Within the same genera there is considerable difference in the quality of wood produced by the different species. Longleaf pine, for instance, is dense and hard and resists longitudinal pressure almost as well as white oak. Loblolly pine, on the other hand, is quite brittle and is of little use where great strength is required.

Rapidity of growth also influences the strength and quality

of wood. As a rule, it may be said that in ring porous woods, like chestnut and oak, the more rapid the growth the stronger is the wood produced, because after spring wood has been formed, the dense layer of summer wood is added to the growth ring. This latter part is really the supporting portion of the ring and the faster the growth, the larger is the proportion of the summer wood. Consequently chestnut sprouts are considered stronger than the more slowly growing seedlings and are in greater demand for piles and poles.

The influence of the quality of the soil on tree growth also is quite remarkable. Light sandy soil tends to produce a fine grained wood which is usually less tough and of less fuel value and durability than specimens grown on heavy soil. On clay soils the growth is apt to be slow in youth and the stands mature later on such situations. The wood is exactly the reverse of that grown upon sandy soil, being heavy, durable, tough, and of good fuel value.

Reproduction. — This characteristic of a tree is of extreme importance to a forester as on the successful starting of new growth depends the future forest crop.

There are three methods of natural regeneration:

1. By seed.
2. By sprouts.
3. By suckers.

1. Seed production depends physiologically on the ability of a plant to elaborate more food than it needs for growth, and after accumulating it in the parenchymatous tissues for several years, it disposes of the extra energy by bearing an unusual quantity of seed. Some seed is produced every year as a rule, but these years of large seed production called "seed years," vary with different species as far as frequency is concerned and indeed within the species, depending on climatic conditions, site, etc.

Some species are noted for the frequency of their seed years and the total amount of seed produced. Others may produce a very small amount or an average amount with low fertility per cent. Light seeded intolerant trees are more likely to produce a great deal of seed and such trees as birch and aspen are liable to seize upon any piece of land whose surface meets their requirements for a germinating bed. If a species makes but slight demands on the soil during the early part of its life, the chances are so much the better that a permanent stand will be established.

Light, next to food of course, is the most important consideration in seed manufacture. Consequently open grown trees produce a larger amount of seed than those grown in close stand and can produce seed at a much earlier age. Twenty-five years is the average age at which open grown trees begin to bear seed, while those in close stands will begin to bear at 40 years as a rule. Forest grown trees bear their seed chiefly at the top, while those submerged beneath the canopy of the forest rarely bear any seed at all. A tree growing in the open with plenty of light available will bear seed at the ends of practically all the branches.

Heat is also of importance in seed production, as a tree growing in its optimum range or toward the southern part of its range is apt to bear more and better seed than a tree of the same species much farther north owing to more favorable growing conditions. Seed years are due to the favorable temperature and moisture conditions of the previous season or seasons and therefore are likely to become more frequent with good climatic conditions and freedom from any disturbance, such as defoliation, drought, etc. They do not occur in the same year throughout the range of the species but the interval may be about the same. For instance, the white pines in Maine may have a splendid crop of cones, while those in the Adirondacks may be bare the same year.

Some species bear seed at a very early age but, as a rule, the seed produced is apt to be of low fertility. Lodgepole pine has been known to produce seed as early as six to eight years; jack and scrub pine at about the same age and pitch pine at about ten years of age.

Seed production, however, is at its best in the economy of the tree after the period of most rapid height growth is past. At this time the crown commences to develop, and the food supply begins to catch up with the demand. When silvicultural maturity approaches, that is, when the stand begins to open up and reproduction starts naturally, seed production is apt to be at its best.

Seed is produced by trees until they are very old and it is not uncommon to find a tree producing a heavy crop of seed a year or two before it dies.

There is a wide difference of opinion about the quality of seed produced by these old trees and the quality of reproduction that will result if they are left as seed trees. Some believe that the seedlings will be weak while others state that while the fertility may be low, the seeds that are fertile will produce good stock as the individual seed bears the imprint of 10,000 ancestors and is not affected by the condition of the immediate parent.

2. *Reproduction by Sprouts.* — Starting a forest by means of sprouts is known as coppicing and is a very desirable means of regeneration with certain species, as maple, chestnut, catalpa, etc. It should not be continued too long, as it is apt to exhaust any but the richest and heaviest soils, on account of the frequent exposure to sun and wind.

Practically all hardwoods sprout during youth but some retain their sprouting capacity longer than others. Seedling chestnut will sprout up to 120 years of age, while sprout chestnut has difficulty in coppicing after 60 years. Under

Fig. 6. — Butt of Redwood Sprout Showing Bark, Sapwood, and Heartwood.

Redwood is one of the few conifers that is capable of reproducing by sprouting from the stump.

ordinary circumstances the coppice method should be used only on short rotations as the sprouting capacity is apt to be uncertain after 40 years, as a rule.

Repeated coppicing tends to lower the vitality of the parent stump, and consequently many blank spots are found in a forest that has been repeatedly coppiced.

A few of the conifers sprout, but the sequoias are about the only species of conifers which produce sprouts of economic importance. Shortleaf and pitch pine are quite frequent sprouters under certain circumstances but the sprouts produced are short-lived and of little importance.

3. *Reproduction by Suckers.* — By suckering is meant the sending up shoots from underground roots. This method of reproduction is not of economic importance, as a rule, because the suckers of most species are extremely apt to die. In certain species, however, such as beech and poplar, growth from suckers is often responsible for the extension and maintenance of the stand. Black locust also reproduces in this way, and when it is once established in a field the more the mature trees are cut out, the denser will be the thicket of suckers sent up from the roots.

Resistance. — Trees, like individuals, are damaged by various agencies, and different species are able to resist these agencies in varying degrees.

Among the enemies to the forest, wind, snow, insects, fungi and fire may be mentioned and each will be considered in detail in the Chapter on Forest Protection.

The individual resistance of a given species is of extreme importance in forest management and the desirability of a tree depends on its vitality and ability to throw off the attacks of various enemies. Trees like white oak, which is particularly favorable to the spread of the gypsy moth, and the white pine, which is badly attacked by the pine weevil and blister rust, are discriminated against in regions where such pests prevail.

An intimate knowledge of the relative resistance of the

different species is part of the forester's stock in trade, as on such knowledge is based the decision concerning mixtures to use in planting or which species should be favored and which should be discriminated against, in a thinning or reproduction cutting.

The Tree as a Unit.

The tree to the forester is the fundamental unit which in large numbers forms the community, the forest, in which he is chiefly interested. It is upon the forest, its relations and reactions that the forester bends his energies. The characteristics of a tree, its soil requirements, its reactions to light, its height, growth, etc., are only interesting and important in so far as they affect the selection of site, the density of the stand, the outcome of the struggle for existence. In other words forestry is a science in so far as it is based upon the close and intimate knowledge of the single tree as a unit; it is an art because it applies this scientific knowledge of the individual, in producing timber from trees grown in groups or communities.

Bibliography

MARSH, GEO. P. The Earth as Modified by Human Action.

MERRIMAN, C. H. Bulletin U. S. Biological Survey.

NISBET, JOHN. Studies in Forestry.

PINCHOT, GIFFORD. Primer of Forestry, parts 1 and 2 published as Farmers' Bulletins 173 and 358. U. S. Dept. of Agriculture.

SCHLICH, WILLIAM. Manual of Forestry. Silviculture. Vol. 2.

SCHIMPER, DR. A. F. W. Plant Geography.

CHAPTER III.

SILVICS.

The Forest as a Community.

A forest is a community of individuals which itself has a life. It is more than a mere group of trees since it includes not only the tree but the soil, the undergrowth and reproduction. The Germans even include the game and stock within the realm of the forester's activity.

The component parts of a forest bear the same relation to each other as do the individuals in a town or village. The trees are mutually dependent and at the same time competitive. They protect each other against windthrow and snow break; the close canopy allows the formation of humus, and yet they compete for food, moisture and light. So sharp is this competition that a stand of 100 trees per acre at 200 years may be the final survivors of 10,000 seedlings which started on that acre, the balance having perished in the struggle for existence.

The duty of the forester is to learn the habits of the trees which make up this community so that congenial species may be encouraged and the maximum growth of the forest as a whole be secured.

Characteristics of a Forest.

The knowledge of the life, development and need of the various trees constitutes the science of silvics, while silviculture, based upon this knowledge, is the art of tending the forest community to produce the maximum amount of timber in the shortest time.

A forest must possess the following characteristics:

First. The crowns *must meet* so as to produce a certain amount of shade.

Second. Natural pruning must have commenced so that there is some clear space on the forest floor, *i.e.*, the canopy must not begin immediately above the ground.

Third. There must be an accumulation of humus on top of the mineral soil.

Fourth. The form of the individual must be typical of the forest, rather than of open-grown trees.

A young plantation will not meet the above requirements and thus cannot be considered a forest until the crowns meet and some natural pruning takes place. The above conditions cannot be fully met in such situations as the arid Southwest where an open stand is unavoidable, owing to limited moisture.

Division of a Forest. — For the purposes of description and management a forest is often divided into various parts. The terms used to characterize these portions will be briefly defined.

Stand. — A generic term referring to a specific part or the whole forest. It usually needs further qualification, as "even-aged stand," "*dense* stand," etc.

Type. — A part of the forest having a distinct individuality that requires separate treatment. It is the ultimate unit in description and intensive management. Types may be based on composition, age, stand per acre, etc.

Influence of Forest on Locality.

Aside from the importance of forests in producing timber, there are certain influences exerted by forests which make them practically indispensable. In certain regions these indirect influences may be paramount to timber production as in the mountainous regions of France for example and in fact in parts of the United States where water is at a premium. In the former country steep slopes are being sodded and planted on a large scale primarily to assist in controlling the mountain torrents rendered unmanageable by over-cutting subsequent to the French Revolution.

The indirect influences which forests exert are felt in three ways; viz. effect of forests on

1. Climate.
2. Precipitation and run-off.
3. Soil.

Climate. — Whatever influence forests have on climate is somewhat localized. It is well-known that the climate of a wooded area is more equable than that of an open country since the ground is less exposed to insolation, the radiation of heat is checked, and the currents are retarded. This action of the forest is of great value in hot countries and advantage is often taken of it in rendering wind-swept areas more habitable by planting windbreaks. Forests exert the greatest change in the climate during summer, with spring following, next autumn, and the least influence of the forest on climate is felt in winter. Humidity is increased by the presence of large bodies of forest cover, since the relative humidity of forest air may be as much as 10 per cent higher than the humidity in the open spaces near by.

Precipitation and Run-off. — It is very doubtful indeed if the presence of forests has any marked effect on the amount of rainfall. However, experiments carried on by French and Swiss foresters indicate that under certain circumstances large masses of forest cover at high elevation may tend to increase the annual amount of precipitation by chilling the moisture-laden atmosphere.

For the most part, however, this effect of forests is not

conclusively proven and the forests are chiefly considered as retarding the run-off of the water precipitated.

Run-off and Erosion. — Following a rain storm the speed with which water reaches the rivers depends very largely on the condition of the surface of the soil and its porosity, as well as on the slope. As far as heavy spring showers are concerned it is estimated that fully 25 per cent of the rainfall is absorbed by the trunks, branches and leaves, causing this portion of the total precipitation to reach the soil after the bulk of the rain has been soaked up by the humus and litter. Some part of the precipitation is evaporated from the trunk and branches directly into the air, and never reaches the soil at all.

Tree roots also have their part in retarding the run-off, for wherever they enter the soil, basins are formed in which small pools of water collect. These act as miniature storage reservoirs running off to the next basin farther down the hill when the upper one becomes full.

The humus, of course, is one of the most important parts of the forest in affecting run-off owing to its composition and structure. It is extremely hygroscopic and can hold many times its own weight of water, and the relative water-holding power of a water shed covered with dense forest is much superior to that displayed by a steep slope denuded by fire. On the whole, the forest from canopy to humus and roots acts as a great sponge which first must be filled before the excess water will run off in the streams and rivers. Snow lies longer in the spring within the forest and when it melts is largely absorbed. Springs are much more common in forested land than in the open, and brooks and streams are more apt to have continuous flow if the water shed is forest-covered than if it is denuded.

The effect of forest cover is most noticeable in mountain-

FIG. 7. — ERODED HILLSIDE, MADISON COUNTY, NORTH CAROLINA.

In regions where soils are heavy and deep, where slopes are steep and rain storms torrential, erosion is all too frequent.

ous countries having steep slopes, deep alluvial soil and with heavy rainfall. Such regions are apt to suffer badly from erosion after the forests are removed. In extreme cases the damage is not only confined to cutting away fields on the upper slopes, but in many cases the bottom lands during floods may be covered with sand and gravel, rendering fertile land unsuitable for agriculture. It is estimated that upward of 200 square miles in the United States are annually laid waste by erosion, while the alluvial soil deposited at the delta of the Mississippi represents the fertility of thousands of acres. Steep slopes subject to erosion should not be clear cut or if some clearing is necessary for agriculture, strips of forest land should alternate with the cleared areas to check the force of the run-off.

Soil. — While trees undoubtedly withdraw fertility from the soil, their general effect is to enrich rather than exhaust. The reason being that the roots of the trees draw large quantities of soluble salts from the lower strata of the subsoil and deposit these salts in the leaves and twigs during the process of transpiration. When the leaves fall these salts are deposited on top of the surface soil and gradually decompose under the action of bacteria, fungi, etc., thereby adding to the richness of the surface soil.

The physical condition of the soil is also improved by the forest. The presence of roots either actively growing or decaying adds to the porosity of the soil, and the effect of humus in correcting the binding quality of clay and improving the cohesiveness of sandy soils is well known.

Nowhere is the influence of forests on soil more marked than in the pine barrens of the South. Numerous cases can be cited where fields were covered by "old-field pine" (pinus taeda), after the Civil War and a splendid crop of timber was grown. On clearing the land splendid crops can be raised for the first three or four years, drawing upon the richness which had been deposited by the forest. At the end of that time the accumulation of humus and organic matter has been exhausted, crops commence to decrease in value and within a few years the land is abandoned and the cycle starts again.

Influence of Locality on Forests.

(Factors of Location.)

A tree or a forest composed of trees is largely affected by environment and consequently ecology, the science dealing with the dependence and adaptation of plants to surrounding factors, is extremely important in the proper handling of forests. Different species react differently to the various factors and as a result different kinds of tree communities are produced.

The most important ecological factors are as follows: Air, light, heat, moisture, soil and exposure.

Air. — Air is composed chiefly of nitrogen and oxygen, and a limited amount of carbon dioxide, all of which are necessary to plant life. The relative proportion of the component parts varies but slightly at different locations and elevations. However, in some regions, the atmosphere may contain certain gases or constituent parts which may have injurious effects on plant life. Sulphur dioxide, for instance, may be found in perceptible quantities in the air near smelting or pulp plants and may kill plant life within a radius of several miles, and salt winds from the ocean are found to be injurious to certain kinds of trees.

All plants take carbon dioxide from the air and in the presence of chlorophyll build up the carbohydrates with the addition of the water taken in by the roots. Some of the legumes have the additional power of absorbing the nitrogen from the air and fixing it in the soil through the medium of nodules of bacteria attached to the roots.

Light. — Light is the source of all energy and has been discussed in its bearing upon growth. The effect of light depends largely upon the season, as its intensity decreases with the obliquity of the rays of the sun. Latitude and altitude also affect light, the intensity decreasing toward the north but increasing with altitude.

The duration of light is as important as its intensity since a definite amount of energy and plant food is required by each species. Those possessing more sensitive chlorophyll or having leaves with better arrangement possess an advantage, as with the same amount of light they can exceed their rivals in vitality. Shade bearers, for instance, can produce vigorous buds with only diffused light, and certain of the tolerant trees, like spruce and beech, can thrive with little direct sunlight. Very little light is required in extreme youth as the seed can germinate in shade and grow for a short time on the food stored in the cotyledons, but growth cannot be kept up for any length of time without the energy which light affords. During advanced age also the tree needs but little light, the greatest demand for light being made during the period of most rapid growth when large amounts of energy are required.

Heat. — For each plant there is a certain optimum, maximum and minimum temperature. Cell division, transpiration and assimilation all require a certain amount of heat. Heat is influenced by the following factors:

First. Elevation above the sea level. It is claimed that in the Alps between 300 and 400 feet increase in elevation results in lowering the average annual temperature one degree.

Second. The presence of large bodies of water. Large lakes or bays store up heat during the summer time and release it slowly in the fall and winter. On the whole their tendency is to retard the seasons and to prevent rapid fluctuations in temperature.

Third. Aspect and gradient. Sites facing southwest to southeast have a higher annual temperature than those facing northwest and northeast and the angle at which the sun's rays strike the ground also affects the amount of heat available.

Fourth. Presence or absence of forest cover. Open countries without lakes and forests are subject to much more rapid change of temperature than the forest regions which are well watered. The forest protects the soil from insolation, increases the relative humidity of the air, retards drying winds and in general renders the climate more equable.

Moisture. — As has been stated in the preceding chapter, moisture is of extreme importance for plant life and growth. There are two kinds of moisture, soil and atmospheric. Some species, like hemlock, prefer locations where there is plenty of both, while others, like the eastern larch, seem to require plenty of soil moisture alone. Atmospheric moisture, however, is of great importance since it is not only the original source of the precipitation, but also it controls the rate of transpiration which, in turn, has its effect upon the inflow of nutrient solutions.

The importance of soil moisture in plant economy is more apparent, having the following functions:

First. It regulates the temperature of the soil by evaporation.

Second. It is the solvent for salts which are absorbed by the tree.

Third. With carbon dioxide absorbed by the leaves it is worked up into starches and sugars in the presence of chlorophyll.

Moisture in the plant has the following functions: it increases the toughness of the wood, it makes the salts within the cell soluble, and permits the plant to accommodate itself to changes in temperature by increasing or decreasing the amount of water evaporated. In locations where the temperature is the same, the amount of moisture available controls, to a large degree, the local distribution of given species.

Plenty of moisture in the soil, however, does not necessarily mean sufficient moisture, since certain factors influence imbibition. Salt in the soil, for instance, has a drying influence on the roots, owing to disturbance of osmotic pressure. Humus acids prevent the taking in of water, and the frozen soil makes it impossible for the water to pass through the cell wall. Consequently a tree growing at high elevations in a rare atmosphere, where transpiration is intense, is apt to suffer from limited water supply in spite of heavy precipitation.

Trees requiring moisture:	Trees enduring drought:
Cottonwood	Norway pine
Beech	Jack pine
Black ash	Pitch pine
Maple	Red cedar
Cypress	Black oak
Spruce	Chestnut oak
Eastern hemlock	Black locust
Eastern larch	Mesquite
Western red cedar	Western yellow pine

The coniferous trees require less moisture than broadleaf species and in some respects might be classed with the intermediate, like chestnut, red and white oak, etc.

Soil. — The character of tree growth in any region is largely a matter of soil and by soil the forester considers not only the mineral portion resulting from weathering of rock, but also the organic material which has been added through the decay of vegetable growth on top of the mineral soil, because on the effect of both together depends the reaction of the tree to the site.

The component parts of forest soil may be described as follows:

Fig. 8. — Vertical Section of Forest Soil near Washington, D. C.

Forest floor, humus, masses of superficial fibrous roots and soil proper are all shown.

The "forest floor" lies on the top of the mineral soil in three distinct layers. The upper layer consists of the recently fallen leaves with little or no change in structure. Beneath this we find partly deccmposed matter, the leaves still showing traces of their original structure; and directly on top of the mineral soil is the humus proper. It is black and crumbly material, showing no traces of its former structure. It is this portion of the forest floor which supplies the richness for tree growth, and the rapidity of the forest growth depends largely on the proper decomposition of this layer.

The requirements for normal decomposition of humus are bacteria or fungi, sufficient heat and moisture, and proper amount of air. Long winters with resulting short growing seasons ordinarily mean deep layers of humus and in the north woods it is not uncommon to find the duff or litter anywhere from one to three feet in depth. Compact soils ordinarily have a deep layer of litter, while porous soils are apt to be extremely hungry and rapidly consume the humus. It takes two to three years for hardwood leaves to disintegrate to humus, while coniferous needles take longer, four to five years on the average and even up to eight years. The extra durability of the needles is due to the presence of resin.

The physical properties of the soil are perhaps of greater importance to the forester than the chemical, because on account of the low fertility required by forest growth practically any soil is able to support tree growth, provided sufficient moisture is available. The chief physical properties of the soil may be enumerated as:

1st. Consistency.
2d. Water-retaining capacity.
3rd. Permeability.
4th. Capacity to become heated.

Of these consistency and incidentally permeability are perhaps the most important. Unless the soil structures are sufficiently permeable, the water will not sink into the soil and, consequently, will not be available when needed. If air is excluded from the soil, it will result in the killing of the roots, since the roots require air as much as leaves. This may be proven by noting the frequency with which trees die when their roots are covered with additional earth following grading operations. This has led to the hypothesis that the knees of the cypress have been developed as a breathing organ to supply air to the roots during the gradual lowering of the southern coastal plain.

The ordinary texture and permeability of the soil is wonderfully improved by the addition of humus. It tends to loosen binding soils, makes them more permeable to air and water, and also adds to the hygroscopicity and water-holding capacity of sandy soils. Humus tends to correct the faults of all soils due to physical structure and at the same time adds greatly to their fertility.

The soil itself has life, as it shows changes in form and composition from time to time. Darwin, as a result of his investigations, has shown the extremely important role played by worms in changing the chemical composition and physical structure of soil, proving that they neutralize acid and assist in working the soil and mould together, and increase porosity.

In addition to animal life in the soil, the bacterial action is extremely important. Investigators claim that one pound of humus contains anywhere from 90 to 250 millions of bacteria and their presence may determine the success of tree growth as well as that of agricultural crops. Spruce, for instance, can grow on poor sandy soil, provided Scotch pine is mixed with it, since the additional nitrogen added to the soil by the mycorrhizae of the Scotch pine roots makes the growth of spruce possible. Animal life is found to be most numerous in the upper two feet of forest soil; below six feet it practically ceases to exist.

Exposure. — The direction which a slope faces often has an important bearing on the kind and quality of tree growth found upon it.

A northern slope gets no full sunlight and the rays fall obliquely. Winds are cool but not drying. Snow melts slowly in the spring, hence it is moister. All points are in favor of easy germination of seed, hence natural regeneration is easy. It is a most desirable aspect.

The east slope is cool and moist; fairly late in spring, vegetation is late in starting, consequently, there is little damage from late frosts. On the whole it ranks next to a north slope.

The south and west slopes are ordinarily the least desirable. The ground is parched by the sun and wind; seeds are apt to dry out, making natural regeneration difficult, and sunscald is quite common in thin-barked trees.

All these conditions may vary in different parts of the country, depending on local factors. In the Puget Sound country, for instance, the west slopes receive the heaviest rainfall and support the densest growth.

Formation of Forest Types.

By forest type is meant the ultimate subdivision of a forest sufficiently distinct to affect its management. Forest types in nature are formed in many ways.

The average forest, if left absolutely undisturbed by natural agencies, including fire, insects and wind, and if secure from lumbering operations, would in the course of a century or more develop into what is called a climax or ultimate type of forest which is suited to the particular region in which it grows. In this forest would be found trees of many ages and many different species, and, for the most part, reproduction would be started beneath holes in the canopy made by the death of veterans. In short, in the virgin forest growth is generally balanced by decay.

Aside from species and composition, types may be separated on the basis of topographic situation, as cove, valley or ridge type.

Age and quality of site are sometimes used as qualifying factors, aside from composition, and may separate portions of the forest containing the same species. Examples of these types would be Northern Hardwoods, Quality II, or White Pine, 21 to 40 years.

Pure versus Mixed Forests. — The controversy over the advantage of growing forests composed of one species alone or several species mixed together has been long drawn out among European foresters. Among American foresters also there has been a wide difference of opinion.

With intolerant species, like western or southern yellow pine, whose natural tendency is to grow alone, there can be little choice unless underplanting is resorted to later, to protect the soil and thereby improve the growth rate. Tolerant species growing at first in a mixed forest may ultimately become pure by crowding out the competing species.

Other species, however, like white pine or spruce, may be grown either pure or mixed with certain advantages attached to either method.

Pure forests are easy to start artificially or naturally during a good seed year. Since the silvical characteristics of at least 80 per cent of the trees is the same, a pure forest is easy to tend. At maturity the marketing is comparatively easy, since there is a large quantity of one kind of lumber to be sold. However, pure forests, composed of 80 per cent or more of the same species, are subject to excessive damage from insects or fungi, and in case of shallow-rooted species may suffer severely from windthrow.

Mixed forests, on the other hand, are not troubled so severely with insects or fungi as there is not such a large number of the required host. Windfirm species protect those with shallow roots and there is less competition, owing to the different heights and the different root habits of the various species. Owing to the dense canopy the soil is kept in splendid condition, natural pruning is more rapid and larger yields are the rule.

However, great skill is needed to bring a mixed forest to maturity. There are many different requirements for light, soil and moisture, and in some cases the tree desired may be crowded out by some faster-growing species of low technical value which was merely inserted as a filler.

One of the most serious drawbacks to mixed forests is the difficulty experienced in marketing small quantities of several different species. While there is no serious drawback in Europe where timber markets are fairly stable, it may prove a disadvantage in the United States. The general tendency in American forests at the present time is to use simple mixtures, preferably in groups, suiting each species to the proper soil.

Life History of a Forest.

Various points covering the development of a seedling have already been given in this and previous chapters. The development of the forest, as a whole, will now be considered.

The simplest condition would be to trace the development of a plantation:

At first the trees stand far apart, each having sufficient room and all crowns reach the ground. At the end of eight to ten years the lower parts of the crowns have developed to such an extent that some of the branches interlace; some twigs die out from too much shade, and natural pruning commences. At this point the life of the forest begins.

The form of the individual tree now changes, the canopy rises above the ground owing to continued natural pruning, leaves accumulate on the soil, the forest floor commences to form and the characteristics of a forest tree become more apparent.

From this time on the struggle for existence is most keen. Tree classes soon form and those possessing the most vigor soon commence to surpass and suppress their less vigorous neighbors.

At the end of, say, forty years the various tree classes are well defined. They may be described as follows:

1. Dominant trees; those overtopping all others.
2. Co-dominant; thrifty trees slightly below class 1, but still receiving light on the sides of the crown.
3. Intermediate; trees rapidly falling behind and now receiving light only on the top of their crowns.
4. Suppressed; individuals hopelessly submerged beneath the canopy.
5. Dead; those trees which have already succumbed in the struggle for existence.

From middle age to maturity the struggle is not so keen as during the period of most rapid growth, but individuals drop out from time to time so that at ninety to one hundred years only a fraction of the number remain which were present at the beginning.

As maturity approaches and each surviving tree has appropriated all the growing space it needs, a still further diminution in numbers is noticed. The canopy opens up, seedlings appear on the ground beneath, and silvicultural maturity, the time when the forest naturally reproduces itself, is at hand.

From this time on the veterans will gradually die and their places will be taken by seedlings growing up from beneath until gradually the ultimate or climax forest for that region and site will cover the ground.

FIG. 9. — TWO-STORIED FOREST. SOUTHERN YELLOW PINE.

Young reproduction has seeded in so uniformly as to form a juvenile forest beneath the broken canopy of the older trees.

Forest Description.

In order to cover all points needed to determine the silvicultural treatment of a forest, all factors of site and every point regarding the condition of the forest which would affect its growth, development or needs must be taken into consideration. To that end the forester makes an exhaustive study of a piece of woodland prior to commencing operations and the following headings are covered:

General Description of Forest.

First. Location and Area. — It is always wise to bound a forest and locate the different portions even though the region be entirely familiar. Accurate measurement of boundaries and estimate of area are essential where lumber is valuable.

FIG. 10. — DENSE REPRODUCTION OF DOUGLAS FIR. COLUMBIA NATIONAL FOREST, WASHINGTON.

Fires by removing the thick layer of organic matter may leave the ground in splendid condition for reproduction provided all of the seed trees have not been destroyed. This timber was fire-killed nine years previously.

Second. Physiographic Features. — Under this heading we include the topographic formation, underlying rock, the soil with regard to general character distribution, and its influence on forest types.

Third. The Forest Proper. — Under this we are concerned with the general composition and origin of the forest whether, for example, hardwood or coniferous, and whether of sprout or of seedling origin.

Under the forest types we consider in detail the following:

(*a*) The proportion of the forest area in each type;

(*b*) The relation between type and site;

(*c*) The species present and the various percentages;

(*d*) The density, age and distribution of forest types;

(*e*) The size of the trees, height and diameter;

(*f*) The condition of the forest floor including humus, litter and ground cover;

(*g*) Character of reproduction.

Fourth. Condition of the forest by individual types. — Under this heading is covered:

(*a*) The silvicultural needs of each type;

(*b*) The damage caused by fire, windthrow, insects and fungi;

(*c*) Possibilities of lumbering and

(*d*) Merchantable condition of the forest.

These topics when properly treated will give a clear and comprehensive idea of the forest conditions that will enable proper recommendations to be given for its management. While some of the points may seem rather minute, the bearing of them all is important and too much rather than too little information is preferable.

Bibliography

HALL, W. L. & Maxwell, H., Surface Conditions & Streamflow. U. S. Forest Service Circular 176.

MARCH, GEO. P. The Earth as Modified by Human Action.

NISBIT, JOHN. Studies in Forestry.

PINCHOT, GIFFORD. Primer of Forestry, Parts I and II, published as Farmer's Bulletins 173 and 358. U. S. Dept. of Agric.

SCHLICH, WILLIAM. Manual of Forestry. Silviculture. Vol. II.

ZON, RAPHAEL. Principles Involved in Determining Forest Types. Forestry Quarterly, Vol. VI, No. 3, p. 263.

ZON, RAPHAEL & GRAVES, H. S., Light in Relation to Tree Growth. U. S. Forest Service Bulletin 92.

CHAPTER IV.

SILVICULTURAL SYSTEMS OF MANAGEMENT.

Silviculture is the art of reproducing and maintaining forests to get the best possible returns. The method of applying this art in securing continuous crops of timber is called a silvicultural system of management.

Forests may be reproduced either naturally or artificially. Artificial regeneration by means of planting, sowing, etc., is covered separately under Chapter VI.

The objects of silviculture are to produce continuous crops of the best timber in the shortest length of time and with the least consistent expense. The objects, however, may differ with the kind of ownership. For instance, private owners will naturally desire the best financial returns as a result of their forestry practice and have this one object in view, whereas public forests have a much broader function. Besides the yield of direct revenues in the form of saw logs, poles, ties, cordwood, etc., public forests, such as our State and National Forests, serve a very important purpose in equalizing the stream flow to prevent erosion and floods and in furnishing a steady supply of water for irrigation, water power, and municipal water supplies. Some forests, called protection forests, are maintained solely for these latter purposes. It is obviously impossible to measure in terms of dollars and cents the value of these functions. For this reason, too, the states and federal government can afford to practice silviculture on the high mountains and poorer soils and leave the better soils for private interests which must secure attractive financial returns from their investment as in any other business.

Trees vary so much in their characteristics and methods of growth, as explained in the chapter on silvics, that various systems have been devised to suit these differences in securing natural reproduction. Some trees, like yellow poplar and western yellow pine, demand full light for growth; others like beech, spruce and hard maple, flourish in partial shade; some scatter their seeds readily in the wind; others grow in even-aged stands or in all-aged stands; hardwoods sprout up from the stump when cut, and so on. Each individual tree has certain characteristics which must be studied and determined before it can be decided to which system of reproduction it is best suited. Altogether the principal silvical characteristics which determine the method to be used in connection with it are: (1) Relative ability to produce and scatter seeds. (2) Sprouting capacity. (3) Character of roots or windfirmness. (4) Tolerance or ability to endure shade. (5) Habit of growing in even- or all-aged stands. (6) Ability to grow in pure or mixed stands.

Methods of Natural Reproduction.

There are three methods of securing natural reproduction, as follows: (1) By seeds. (2) By coppice or sprouts. (3) By root suckers. Of these, reproduction by seed is by far the most important. Sprout reproduction is only used successfully with certain hardwoods in the East, while suckering is relatively unimportant and is not relied upon to reproduce our forests with any very important species.

By Seeds. — Reproduction directly from seed is secured by leaving so-called mother trees singly, in groups, or in strips to reproduce a cut-over area. Most of our trees begin to produce seed when quite young and only scatter seeds every few years. For instance, Norway pine seeds every three or four years, while white pine seed years occur six or seven years apart. Lodgepole pine begins to produce seed at six years of age and red spruce at fifteen years. The quantity of seed and the frequency of seed years depend on the amount of food material stored up in the trees. Therefore, trees seed less and more infrequently in the northern limits of their distribution, because the conditions of growth there are not so

FIG. 21. — BRUSH PILING IN MINNESOTA.

Brush disposal is sometimes used to aid in natural reproduction by exposing the mineral soil for seed germination. Its chief purpose, however, is usually for fire protection.

favorable. Light is the great single factor in seed production. Every one knows how full a chestnut or pine is with seed when growing in the open. In dense stands, trees produce very little seed because each individual tree gets very little light. Trees under shade seldom seed. The best seed is produced when the energy of the tree is greatest. This accounts for the fact that some seeds have a high percentage of germination, because they are taken from vigorous, healthy trees. Therefore, trees left for seed production should be young or middle aged, rather than old, defective specimens.

Seeds are dispersed in many different ways. Some are light; others are heavy. This is important in natural reproduction, because on it depends whether a system of clear cutting or gradual thinning can be adopted in renewing the stand. The principal means of distribution are wind, birds, animals, water and gravity. Willow and yellow poplar seeds are light and are often carried a mile or more from the parent tree. White pine seed has been blown one-half mile and given good reproduction. Heavy seed, such as the chestnut, hickory and walnut, are scattered by small animals, such as chipmunks, squirrels, etc., and by rolling down slopes by gravity. Red cedar and cherry are examples of species distributed by birds. Water often carries seeds great distances. Cottonwood and willow on the sand bars and banks of our rivers are often started from seeds carried downstream.

Many seeds are scattered, of course, that never germinate. It is estimated that millions of seeds are sown that never spring up, because insects and rodents destroy them or because they do not reach a favorable germinating bed. Some trees prefer a leaf mould or old rotten logs on which to germinate. This is especially true of hemlock and spruce. Others are partial to a mineral seed bed, as in the case of Douglas fir and white pine. Some trees, such as lodgepole pine and Jack pine, require a great amount of heat to open their cones. The grazing of sheep often stirs up the soil and assists the germination of seeds.

By Coppice or Sprouts. — Sprouting is confined to the broad-leaved species although pitch pine, shortleaf pine and redwood often sprout after cutting or being killed back by fire.

Most hardwoods such as chestnut, the oaks, ash, and maple, sprout from dormant buds in the root collar, at the surface of the ground. Some, however, such as poplars and willows, will often sprout from adventitious buds at the surface of the cut.

Sprouts are shorter lived than trees from seed and after a certain age, usually forty years, their capacity to send out shoots rapidly diminishes. Sprouts grow much faster than trees from seed, because they have reserve energy in the form of food material stored up in the roots. Chestnut is our best sprouter and will often shoot up eight to ten feet the first year after cutting. It will produce telephone poles and cross-ties in forty years and good sized saw logs in fifty years. Sprout stands are usually managed, however, on short rotations up to thirty or forty years to produce fuel, fence posts and rails and other small materials. Catalpa will often shoot up ten to twelve feet in one year when cut back. This gives a tall straight growth. Eucalyptus sprouts have, in exceptional cases, reached a total height of sixty feet in ten years.

By Root Suckers. — Shoots from the roots of trees are called suckers. Some species sucker when apparently in good health while others, such as elm, basswood, and sycamore, send out shoots from the roots only when wounded. Beech and aspen are, perhaps, our best examples of root suckering. In fact, in certain regions reproduction is almost wholly secured by this method. As a general method of reproduction, however, suckering is relatively unimportant and it is doubtful if it will enter into forestry as a means of renewing the forests on a commercial scale. Practically all aspen stands are reproduced by suckering.

Natural Versus Artificial Reproduction.

The question of securing reproduction of forests by natural means or by planting or sowing has always been a very important one in forestry. Much can be said in favor of each method, but the local conditions will be the determinant factor in each case. In Europe, where forestry has been practiced for from three to four hundred years, artificial regeneration is by far the principal method employed, especially where the most intensive practices may be followed. There is, however, a growing tendency even in Germany and France to follow nature more closely. It may be said, nevertheless, that in this country it will be many years before planting will be resorted to on a large scale, because it is so comparatively expensive, costing from \$6 to \$10 per acre at least. In addition, many of our species are adapted to regeneration by natural means either through sprouting from stumps or the distribution of seeds. White pine in the Lake States and Northeast, loblolly pine in the South, lodgepole pine in the Rocky Mountains and Douglas fir in the Northwest are splendid examples of trees that reproduce themselves naturally and successfully from seed. Wherever forests have been denuded by lumbering and then burned by disastrous fires, the only practical means to employ is to plant. New York has already planted about 12,000 acres on state and private lands. The United States Forest Service reforests about 30,000 acres annually by planting and seeding.

Advantages of Natural Reproduction. — The great argument in favor of natural reforestation is its cheapness as compared to artificial methods. The only cost involved is in the leaving of seed trees individually or in groups or strips. These may be a few thrifty growing or old defective trees so that this investment may be almost negligible. Natural reproduction also follows nature more closely and the kind or mixture of trees best suited to the locality is the result or can be easily secured by proper thinnings. There is apt to be less washing of the soil and exposure to the elements, because with most of the methods of natural reproduction the ground is protected by at least a partial forest cover. Then, too, in this country where intensive forestry will not be practiced until our virgin forests are more completely cut over and market conditions improve, natural reproduction can best be practiced especially in our rough mountains and remote forest areas.

Advantages of Artificial Reproduction. — As opposed to the above, the following is a summary of the advantages of establishing and renewing forests by artificial means:

1. Reproduction is secured immediately. Seed years are often far apart and natural reproduction may require a number of years to satisfactorily cover the cut-over area.

2. It is sure, provided proper methods and species suited to the soil and climate are used.

3. The best trees suited to the local market may be used. Sometimes species not naturally growing in the locality may be employed with success.

4. Each tree is given the proper space for growth which in the end means a greater and better product.

The Cost of and Returns from Silvicultural Practice.

In the practice of silviculture, it is obvious that it is more expensive than the old method of lumbering in this country which had no thought of the future. And in order to be practicable it must insure in the long run an adequate return on this increased investment. Perhaps this can be best explained by a comparison of wood production in this country and abroad. Under our present haphazard system of care-

FIG. 12. — MARKING MATURE TREES FOR REMOVAL UNDER THE SELECTION SYSTEM. COCONINO NATIONAL FOREST, ARIZONA.

In the Southwest, western yellow pine grows largely in even-aged groups. The group selection system is, therefore, applied with best results.

less cutting, enormous fire losses and improper care of our growing forests, it is estimated that on an average only twelve cubic feet of wood are produced per acre annually in this country. Contrasted with this the annual production of wood in Germany is over forty-eight cubic feet per acre. In other words, it has been very profitable for Germany to engage in the intensive practice of silviculture. Their economic conditions are responsible in a measure for this, but they do not in themselves explain this wide discrepancy.

The added costs that are usually involved in practicing silviculture may be briefly enumerated as follows:

1. Increased cost of logging in removing only a portion of the stand at a time and in protecting young growth from injury.
2. Cost of marking trees for removal.
3. Value of mother trees left for seeding purposes.
4. Cost of supervision and protection from fire, such as brush burning, patrol, etc.

It is apparent that the cost of these operations will vary greatly according to the silvicultural system employed, and the local conditions, such as the markets, cost of labor, type of forest, method of logging, etc. For a lumberman to employ fairly intensive management, the increased cost would be between 50 cents and $2 per thousand board feet logged. This cost would be distributed about as follows: Fire protection and supervision would cost annually about 2 to 7 cents per acre; brush burning, if used, about 25 to 75 cents per thousand; cost of securing reproduction from $2 to $8 per acre and increased cost of logging from 20 cents to $1 per thousand board feet.

The returns, on the other hand, will also vary greatly. The forests of one of the German states have averaged as high as

$5.40 net per acre per year. Some of the famous municipal forests have yielded $7 per acre per year. In New England, white pine plantations have yielded as high as 6 per cent on the original investment, even under our comparatively unfavorable market conditions. Some Middle West plantations have brought an annual return of over $10 per acre per year with catalpa, white pine and cottonwood. Several of our states are purchasing large areas of cut-over or second growth forests which will no doubt pay good returns as a long time investment. Ordinary forests in this country may be made to yield at least from $1 to $4 per acre per year. This can be done on soils unsuited to agriculture which would otherwise be unproductive waste land. With improved market conditions and better methods of logging, the returns in the future are sure to be even greater.

Silvicultural Systems of Natural Reproduction.

The following systems of reproducing the forest by natural means are generally recognized:

1. Selection system.
2. Clear-cutting systems.
3. Shelterwood system or stand method.
4. Coppice or sprout system.

Each of these systems is carried out in practice with a great many variations to suit the local requirements. Under intensive practice the forest type is the basis of management. The type may extend over a few or many hundred acres. In any case, the whole tract is usually sub-divided into compartments and each compartment is cut over annually or periodically so that there is a continuous yield. For instance, if a tract of white pine is to be managed on a rotation of 100 years and a portion of it is to be cut each year, there will be 100 compartments. Thus, when the last compartment is cut over, the first one will have reached maturity and will be ready for the axe. If, on the other hand, it is only planned to make a cutting every five years, this area would be sub-divided into 20 compartments.

Selection System. — Under the selection system the trees are removed individually or in groups as they reach maturity and there is a gradual succession of young trees growing up to take their places. It is the popular idea of the means that are employed to keep our forests in a continuously productive state. With our rough and extensive system of forestry in this country, the selection system is easily applied and requires less skill than any of the others except the coppice system. Many of our lumber companies employ it in a rough way in taking out of the woods only the larger and finer specimens of merchantable trees. It is the method principally employed up to the present time on our National Forests, because market conditions do not often permit the use of the clear-cutting or shelterwood systems. It is also best used in our woodlots, where the larger trees can be cut from time to time and the smaller ones left to grow up and develop, while seedlings take the places of those removed.

Under the principles of this system, however, it is only applicable to stands where all-age classes are represented (selection forest). That is, if a cut is made every ten years on a rotation of sixty years, there should be at least six distinct age classes represented. This distribution of age classes must be maintained, because only the periodic or annual growth should be cut, and it is necessary to have the requisite number of young trees growing up to fill the places of those removed at the time of cutting. For example, on a tract of 1000 acres if the annual growth of all the trees is found to be 500 board feet per acre, this amount could be removed annually per acre, or 5000 board feet per acre every ten years.

A rough method commonly followed to regulate this yield is the use of the diameter limit. That is, all trees above a certain minimum diameter are removed unless needed for seed purposes or to protect the soil from erosion. This method is especially useful in placing a virgin forest under at least some rough system of sustained yield. In the West lodgepole pine is often cut to a diameter limit of 10 inches; western yellow pine to 15 inches and Engelmann spruce to 14 inches. In the East spruce is cut to 8 inches for pulp wood, hardwoods, such as beech, birch and maple, to 15 inches; hemlock to 14 inches, etc. Certain undesirable species in the stand can easily be discriminated against by reducing the diameter limit for them so that, through repeated cullings, they will be eliminated in favor of better species.

In addition to being applicable to selection forests (stands of all age classes), this system should theoretically be used with tolerant and windfirm species. Where all ages are represented, the young trees will have to come up under partial shade at least and when the large trees are removed it opens up the stand considerably and many trees are likely to be thrown over by the wind unless they are deep and strong rooted.

Whenever a selection forest contains different age groups instead of individuals, tolerance and windfirmness are apparently not so important. Western yellow pine in the Southwest, an intolerant tree, is a good example of management by the group selection method.

The advantages of the selection system may be summed up as follows: 1. It is easily applied to virgin forests and tracts where only rough forestry is permissible and only the largest and best individuals can be marketed at a profit. Little skill is required.

2. It maintains the forest cover and therefore prevents erosion and floods.

3. It is inexpensive. The only added cost is slightly more expensive logging and the cost of marking trees for cutting which is only about from 3 to 10 cents per thousand board feet marked.

FIG. 13. — CLEAR CUTTING WITH INDIVIDUAL SEED TREES. MINNESOTA NATIONAL FOREST.

Ten per cent of the trees in the original stand of white and Norway pine are left to seed up the area.

Clear Cutting Systems. — With this system the forest or a portion of it is cut clear. There are a great many variations of this system. Reproduction may be secured from adjacent forests or from seed trees left standing in the area logged either individually, in groups, or in large blocks. It is principally applied in Europe by cutting consecutive narrow strips each year or periodically against the direction of the prevailing wind.

The conditions that require the use of the clear cutting methods are:

1. Where trees are shallow rooted or grow in exposed places and there is danger of being thrown by the wind. Examples: red spruce and white cedar.

2. Where trees grow in even-aged stands so that logging can be done cheapest by removing the whole stand and whatever reproduction is present cannot be protected. Examples: Douglas fir, redwood and white pine.

3. Where trees are intolerant of shade and need full light for reproduction. Examples: western white pine, western larch and southern cypress.

4. The use of light seeded species, such as pines, spruces, yellow poplar, etc., the seed of which can easily be carried by the wind.

Conspicuous examples of the use of clear cutting in this country are in the Northwest with Douglas fir, lodgepole pine in Montana, western white pine in northern Idaho and red spruce in the Adirondacks on exposed situations. On the Deerlodge Forest in Montana, lodgepole pine seed trees have been left singly, in groups and in strips. Reproduction with this species is so thrifty when the soil is burned over to open the cones and prevent fire danger that the young trees often come in too dense for best development. On this forest clear cutting is advisable not only on account of the silvical characteristics of lodgepole pine but also on account of the market conditions. The Butte mining district uses not only saw logs and mining stulls but also enormous quantities of mine props and lagging and fuel wood so that there is complete utilization of all available material logged from the forest.

Leaving white pine seed trees scattered over the lumbered area has been tried in Minnesota but the isolated seed trees were blown over by the wind. This system was also unsuccessfully used with western white pine in Idaho but by leaving large blocks or groups of seed trees, for mutual protection, this difficulty was avoided and excellent reproduction secured. These blocks were located on ridges and vantage points to obtain the best seed distribution.

FIG. 14. — CLEAR CUTTING METHOD LEAVING SEED TREES IN GROUPS. WHITE PINE TYPE, NORTHERN IDAHO.

About 25 per cent of the area is left in groups of trees on elevated positions to seed up the area that has just been cut clear. The brush is burned broad cast or in piles.

The yield may be regulated in the same way as explained in connection with the selection system, that is, only the growth is cut from time to time.

The brush and slash after logging is disposed of either by piling and burning the brush or by burning it broadcast.

This is done both to reduce the risk from fires and to expose the mineral soil so that the seeds may germinate more easily.

The disadvantages of any of the adaptations of the clear cutting systems are as follows. The advantages have been briefly mentioned above.

1. Clear cutting exposes the soil and leaves it susceptible to deteriorating influences, such as erosion, baking of the soil, loss of humus and occupation of the soil by grasses, weeds, etc.

2. The young trees are exposed to damage by insects, frost and drying by wind and sun.

3. It is only applicable to light seeded, intolerant trees that grow in even-aged stands, and, under certain conditions, trees that are windfirm.

It may be added, however, that the disadvantages of clear cutting may be minimized by cutting very small areas or narrow strips at one time. The larger the areas cut, the greater are the dangers and likelihood of securing unsatisfactory reproduction.

Shelterwood or Stand Method. — This is a highly theoretical system of management and will not be used, except in a very rough way, for a long time in this country. It is seldom used even in Germany and France and demands the most favorable economic conditions for its successful application.

It is a system of management whereby reproduction is secured by means of a systematic progression of thinnings. It is applicable only to even-aged, tolerant species that are windfirm.

At first the shelterwood system involved the use of only two general thinnings or openings in the stand to secure reproduction. The following three are now used when the system is intensively applied:

FIG. 15. — REPRODUCTION OF NORWAY SPRUCE AND SILVER FIR. VILLENGEN, BLACK FOREST, GERMANY.

The shelterwood system is successfully used with natural reproduction here. This shows the condition of the forest after the reproduction cutting.

1. *Preparatory Cuttings.* These are begun during the middle of the rotation and open up the stand, by the removal of a few trees (from 15 to 35 per cent), to prepare the soil for the germination of the seeds, to stimulate seed production by giving more light to the intended parent trees, and to accustom the remaining trees to more or less isolation and give greater windfirmness. If the stand is dense, this thinning should be made much heavier than in a thin or open stand.

2. *Seed Cutting.* This consists of one or more thinnings a few years later that open the stand still farther and is usually done just before or during a seed year. This removes from 30 to 60 per cent of the remaining trees and has as its

object the opening of the stand so that the young reproduction resulting from this seed year may develop in the spaces and light made by the thinnings.

3. *Removal Cuttings.* All the remaining trees may be removed in one or more cuttings. By this time the reproduction that will form the next rotation is well under way and needs all the light possible for full development.

FIG. 16. — A ROUGH APPLICATION OF THE SHELTERWOOD SYSTEM IN THE BLACK HILLS, SOUTH DAKOTA.

Most of the stand of western yellow pine has been removed. When satisfactory reproduction has been established, the remaining trees will be cut.

These three groups of thinnings or cuttings may be carried on over the whole stand periodically or may be developed gradually over the forest by means of strips or groups.

The shelterwood system is used in a very crude way in the Black Hills with western yellow pine; with eastern red spruce; lodgepole pine and with longleaf pine.

The advantages of the system are as follows:

1. The progressive thinnings leave a large number of seed trees evenly distributed.

2. It does away with most of the disadvantages of the clear cutting systems since the soil and reproduction are not exposed in any way.

3. The new stand is established in advance of the final cutting and the thinnings give the maximum rate of growth to both young and old trees.

4. The system may be used with heavy seeded species.

5. Trees not desired in the new stand can be eliminated in the preparatory cuttings.

The disadvantages are as follows:

1. It requires considerable skill and the best possible market conditions in order to profitably dispose of the product of the thinnings and final cuttings.

2. Logging is very expensive as the forest is worked over so many times in cutting off the trees.

3. Reproduction is exposed to serious injury both in falling the trees, in thinnings and in removing them from the forest along skidways and roads.

4. Theoretically it should not be used where there is danger from windfall or with intolerant species.

Coppice or Sprout System. — This is the simplest method of securing natural reproduction. On account of the excellent sprouting capacity of most of our hardwoods, the forest may be reproduced by merely cutting off the trees at maturity. This system has been used in parts of New Jersey, Pennsylvania, Connecticut and southern New York to supply some of the smaller products of the forest, such as cordwood, poles, posts, ties and mine timbers for many years. The old stumps, however, deteriorate after a certain number of rotations and the forest must be renewed from time to time from seed. This has already happened in some parts of the East. In Europe there is a growing sentiment in favor of the increased use of seedlings in sprout forests. In fact, even in the first rotation of management under this system, most hardwoods lose their maximum sprouting vigor after a certain age. This limit is about 25 years with nearly every species, except chestnut which apparently retains this power for a much longer time. Vigorous chestnut trees have been produced from stumps 120 years of age. Some trees fail to sprout after 40 to 60 years of age. In Europe sprout forests are often cut on rotations of 10 to 20 years to produce fuel wood.

There are certain rules which should be observed in cutting trees to obtain the best sprouts as follows:

1. Trees should be cut between October 1 and April 1, during the period of vegetative inactivity. In the spring, the new shoot will often grow up to a height of 8 to 10 feet. If cut in the summer, sprouts are still weak when winter comes and are likely to be winter killed or seriously injured by the cold.

2. Stumps should be cut low in order to make the sprouts as vigorous as possible. Decaying stumps often infect or interfere with the growth of the new tree. More valuable wood can also be secured by cutting low stumps.

3. Stumps should be cut smoothly and sloping so that water will run off and the danger from insects and fungi

decreased. The bark should also be left intact for the same reason.

4. A given area should be cut clean in order to give the greatest amount of light to the young trees, and brush should be piled away from the stumps.

Variations in sprout management in leaving certain sprouts and seedlings over for a second rotation are sometimes used especially to rejuvenate the stand by increasing the number of trees from seedling origin. When this is not successful, planting must be resorted to.

The region of best sprouting in this country is along the Atlantic coast, east of the Appalachian chain of mountains, from Maryland north to Massachusetts. The best tree for sprout management is undoubtedly chestnut on account of its sprouting capacity, rapid growth and the excellent quality of its wood. Unfortunately the bark disease may eliminate it from future management. Other important trees that sprout to advantage are the oaks, hickory, basswood, yellow poplar, red maple and white ash.

Combination of Silvicultural Systems.

It is seldom that the silvicultural systems, depending on natural reproduction from seed, described above, are followed out in this country according to theoretical principles. Very often local conditions, such as markets, topography, type of timber, etc., make many changes necessary. Frequently combinations of various systems are employed. Natural reproduction is often assisted by planting young trees or direct seeding to fill in the open spaces where the expected reproduction has not seeded in naturally. In Europe, in fact, planting is resorted to in the majority of cases where clear cutting is used.

Factors Governing the Choice of Species for Management.

In choosing the proper trees to favor in the management of any forest tract, there are several considerations which must be kept in mind. These factors are:

1. The most important is rapidity of growth. A certain tree may yield a wood of high technical qualities but if it requires 200 or 300 years to reach merchantable size this will discredit it. Some of our redwoods are three thousand years old. Longleaf pine and southern cypress, although producing excellent woods, are too slow in growth. Chestnut, red oak, Douglas fir and white pine are rapid growers.

2. Quality of wood is apparently very important. If two trees of equal rapidity of growth are of unequal demand on the market, the better one should be favored.

3. Reproductive ability is important. A thrifty and easy seeder should be preferred to a tree of infrequent seed years, low per cent of germination and one whose seed is distributed with difficulty. Some trees are also planted much easier than others on account of their small root system.

4. Silvicultural value is also a consideration. Some trees help the growth of others in association with them and improve the quality of the soil. Beech, for example, is sometimes called the " mother of forests."

5. Freedom from injury by wind or attack by insects and fungi. In some localities the borer attacks the black locust so seriously that this species cannot be encouraged with safety. The example of the chestnut blight is also self-evident.

6. Suitability to the local market is important. For instance where mine timbers or cooperage stock or any other special forest product offers a good market, it may be advisable to grow trees that will be most suitable for this demand.

Altogether some of our best native trees for encouragement in American silvicultural practice, based on the above considerations, are white pine, Norway pine, Douglas fir, loblolly pine, shortleaf pine, western white pine, western yellow pine, sugar pine, chestnut, red oak, basswood, yellow poplar, white ash and cottonwood. Some of the best exotic species for use in this country are Scotch pine, Norway spruce, European larch and eucalyptus.

BIBLIOGRAPHY.

GRAVES, H. S. Principles of Handling Woodlands. John Wiley and Sons, New York City. Pages 1–188.

GRAVES, H. S. The Selection System. Proc. Soc. of Am. Foresters. Vol. V, No. 1, 1910.

GRAVES, H. S. Condition of American Silviculture. Proc. Soc. of Am. Foresters. Vol. 3, No. 1, 1908.

HAWLEY AND HAWES. Forestry in New England. John Wiley and Sons, New York City. Pages 17–36.

MOORE, S. L. Regulating the Cut of the National Forests. Proc. Soc. Am. Foresters. Vol. V, No. 1, 1910.

SCHLICH. Manual of Forestry. Vol. II, Siliviculture. Bradley, Agnew, and Co., London. Pages 93–119.

CHAPTER V.

IMPROVEMENT CUTTINGS.

Reasons for Improvement Work.

While nature unassisted provides splendid stands of timber after a lapse of centuries, economic forestry requires maximum production of timber in the least possible time and with minimum expense. Consequently, all measures should be taken that will improve the quality of timber, shorten the rotation and diminish the necessary charges, considering the forest soil and crop as an investment. With these ends in view, certain improvement work is imperative under average conditions, since the expenditure of a little money will often prove a splendid investment, by largely increasing the returns. The removal of undesirable species and weaklings with consequent diminution of competition, results in increasing the volume growth and vastly increasing the quality and yield.

Kinds of Improvement Work.

Cuttings made in immature stands of timber for the purpose of accelerating growth, diminishing competition, or improving the mixture, are of prime importance. The starting of new growth is not aimed at primarily in making these cuttings, although a heavy thinning in a forest old enough to produce seed may often result in a dense stand of seedlings beneath.

Improvement cuttings may be classified under four heads:

1. Cleanings;
2. Liberation Cuttings;
3. Thinnings;
4. Damage Cuttings.

Cleanings. — A cleaning is a cutting made in a very young stand to improve the mixture and reduce competition. Such an operation hardly ever yields a profit, but ordinarily requires the expenditure of from 50c. to $2.50 per acre. Therefore, a cleaning is warranted only under conditions where intensive forestry may be practised.

Cleanings are needed where desirable species are being badly crowded or suppressed by fast-growing forest weeds, and the object is to keep the desired species in the mixture by cutting back the weeds and releasing the slower-growing seedlings which will ultimately yield timber of higher technical value. Such a condition is often found in the young sprout stands where species like basswood, ash, red oak, etc., are being rapidly choked out by such strong sprouting species as black oak, aspen, maple, etc. In such circumstances, a few hours' work per acre may result in a vastly improved stand of timber. The seedlings in each case are favored at the expense of the sprouts.

In many New England pastures which have seeded in to gray birch, pine or spruce often come in under the nurse trees. As the stand grows older, the injurious effects of the birch become more and more apparent. At the beginning, the light shade was an advantage in preventing the drying out of the soil, but as time passes, the top shoots of the pine and spruce become enmeshed in the branches of the birch, and it is very common to have the leaders and side branches of the conifers badly damaged by the whipping action of the birch. This whipping action may be so marked that large holes will surround each birch in a dense stand of pine. In such circumstances, a cleaning during the early life of the mixture would have improved the stand and greatly reduced this damage.

Cleanings are also required where a plantation has been

made beneath dense underbrush such as a stand of sumach, or scrub oak. In the scrub oak country of Pennsylvania, or Long Island, where the ground is practically in possession of this tree, it would be necessary to clean the land before planting, and then carefully watch the plantations to see that the seedlings were not crowded out. A brush hook or light axe is the best tool for cleanings, although in young sprout stands, shears may work well. In some cases, the expense may be decreased by hacking off the top of the undesirable specimens rather than cutting them off flush with the ground.

Cleanings should be made in young stands as soon as the crowding effects are noticed. Usually, in a mixed stand, this will occur during the first ten years. If economic conditions warrant it, a light cleaning should be made which can be repeated within five years.

The most important point regarding cleanings is to decide whether or not conditions will justify the expenditure, and how much should be taken out. When it is realized that in a stand of young sprouts and seedlings, the investment of $2.00 per acre may free 300 desirable seedlings five years old, and that this number will more than fully stock the stand at maturity, it may be seen that cleanings are often highly profitable.

Liberation Cuttings. — Where an abandoned pasture has seeded in, it is quite common to find occasional large trees scattered among the young seedlings forming the majority of the stand. As the seedlings develop, the shade cast by the advance growth becomes more injurious, and the young tree may be throttled by the larger trees, which accounts for the name "wolf tree," coined by the German foresters. In addition, the excessive root competition of these large spreading trees often makes profitable growth of young seedlings impossible within a wide radius. In such circumstances, the removal of this advance growth is necessary, in order that a full stand of straight, desirable saplings should result. Contrary to cleanings, liberation cuttings may be quite profitable, since large trees are removed, and, consequently, good timber may be obtained. The quality, however, is often poor on account of many knots, and the scattered trees make logging costly. A liberation cutting should at least pay expenses.

The trees removed are, for the most part, much older than the seedlings which they are damaging, and this difference in age constitutes a distinguishing feature. For this reason, less skill is required in selecting the individuals to be cut, and liberation cuttings may, as a rule, be carried on independently of other cultural operations.

Thinnings. By thinning is meant the removal of individuals in an immature stand too dense for rapid growth, in order to diminish competition for light, food and moisture. The fastest-growing individuals of the best form are favored and dissipation of growing energy prevented. The result of a thinning is to greatly shorten the struggle for existence, and, by focusing all the growing energy that the site affords upon a few selected individuals, much greater volume growth is obtained, and the quality increment is much higher.

Reasons for Thinnings.

With increasing age, each tree in the stand demands more room both for crown and roots, which results in excessive competition for light and plant food. Under nature's régime, this struggle is prolonged and the trees which have been leaders from the beginning gradually surpass their slower-growing neighbors and eventually appropriate the space which they once occupied. Close planting or a dense stand is necessary for a time in order to cause natural pruning, and indeed, the German foresters insist upon it in order that the early growth shall not be too fast, and, therefore, sacrifice

FIG. 17.—WHITE PINE STAND AFTER THINNING—KEENE CO., NEW HAMPSHIRE.

By reducing the number of stems per acre, excessive competition is reduced and the wood produced is laid on the trunks of the more desirable specimens.

strength. After a few years, however, (ranging from fifteen to twenty-five, depending on the species and site), interference in the form of a thinning is advisable. Even the dominant trees do not grow at their maximum rate, and by relieving the desirable trees from excessive competition, the growth rate is vastly increased.

Results of Thinning.

Improvement cuttings in immature stands have the following beneficial results:

1. They tend to greatly shorten the rotation.

Whereas a stand of sprout chestnut will ordinarily yield a tree ten inches in diameter in forty years if untouched, the same forest, properly thinned, will produce a tree the same size in about thirty-five years.

2. A large amount of material is utilized which would otherwise decay and be a total loss were the usual struggle allowed to proceed.

By utilizing the young saplings for fuel, posts, bucket stock, etc., when they are green, the material can oftentimes be profitably sold, whereas if they were allowed to remain, the quality would be greatly damaged through insect and fungus attacks. The amount of material removed by thinning during the life of the stand may constitute from 20 per cent to 60 per cent of the final yield and this "intermediate yield," so-called, may often determine whether or not the stand itself is profitable. Indeed, in the Duchy of Baden, extremely heavy thinnings are advocated, to reduce the amount of forest capital invested, as a means of increasing the financial profits. The intermediate yield, in this case, is necessary, not because of silvicultural requirements, but because too much timber capital is involved.

3. The removal of the competing trees does not decrease

the final yield, on the contrary, it is increased, considering the increased growth rate. The quality also is greatly improved.

4. By discriminating against undesirable species, the stand at maturity will contain the species in the proportion desired. This is particularly desirable where natural regeneration is practised, as each succeeding rotation requires less cultivation.

5. In stands which have been consistently thinned, windthrow and breakage from wind are lessened owing to the fact that the individual trees have had greater opportunity for root and bole development. Consequently, the individual trees are much stouter and stronger and are able to resist the force of destructive winds.

Severity of Thinnings.

Severity of thinnings largely depends on the product desired as well as the condition of the stand itself. The danger from windfall also may be so excessive that thinnings in older stands must be avoided altogether. Ordinarily, even if the species is not wind-firm, by thinning early and often, a firm root habit may be developed.

If maximum volume production is desired, heavy thinnings are the rule. If timber free from knots and of good technical value is aimed at, moderate thinnings should be made at frequent intervals.

The severity of thinnings is measured on the amount of timber removed, and the size of the holes made in the canopy. It is a safe guide, as a rule, to make no holes in the canopy that the trees cannot fill by crown growth within five years. In some cases this is unavoidable, but if too heavy thinnings are made, the soil is exposed to the drying effects of sun and wind, weeds and grass spring up which exhaust the soil, and instead of increasing growth, the growth rate may ultimately show a decrease. Thus, too heavy thinnings are apt to defeat the end in view. In addition, in the case of hardwoods, increased light may cause the sprouting of adventitious buds along the trunk, producing a brushy bole, and diminishing the amount of clear lumber that may be later obtained.

The severity of thinnings is graded as follows: (See tree classes, Chap. 3).

Grade A. Light, removing dead or dying trees.

Grade B. Moderate, removing all suppressed, and in addition the lower intermediate trees.

Grade C. Heavy, removing in addition the remainder of the intermediate trees.

Grade D. Very heavy, removing in addition many of the co-dominant trees.

A heavier thinning than any of the above is called an accretion cutting. A cutting of this kind is rarely made in American forestry practice, although the same end may be gained by starting reproduction cuttings (see Chapter 4 under Shelterwood System), rather early and permitting the seed trees to profit by the increased light. In some parts of Europe this has been carried to extremes, as the highly profitable growth made by the seed trees led the foresters in charge to hold them over to the injury of the reproduction.

While no hard and fast rule may be laid down concerning thinnings in actual practice, nevertheless, certain methods have been found to yield the best results. During the first part of the rotation, at least, a " C " thinning is necessary, while later, a " D " thinning may be used.

The French system of thinning is considered quite desirable, especially for white pine. According to this method, the trees to form the final crop are selected, and all trees in any way interfering with the selected specimens are removed, regardless of tree class. The remainder of the stand is untouched, except that dead and dying trees are removed. By selecting 150 to 250 stems per acre and favoring them, bigger and better trees can be produced and the soil is protected by retaining the intermediate growth.

Another system of thinning, Borggreve's method, is sometimes used in Europe, which is almost opposite in principle to the preceding. According to this plan, the largest trees are cut, and the dead and suppressed trees besides, leaving the co-dominant trees to furnish the final crop. This has many drawbacks, since large holes are made in the canopy and the trees which are left cannot always recover from earlier suppression, and, consequently, the growth rate does not show the pronounced increase desired.

In American forestry practice, the " C " grade of thinning, after the French fashion, has the widest use, and will ordinarily give the best results.

Damage Cuttings.—It is a decidedly unusual forest that is not visited from time to time by one or more destructive agencies. The western yellow pine of the West has been hard hit by the beetle at various times; the larch saw-fly has inflicted heavy damage upon the tamarack of northern New York, and the Lake States, from time to time. Following the visit of any destructive agency, whether insect, fire, fungus disease or wind, it is only good management to remove the injured material before it is a total loss. In addition, if either insects or fungus disease are the destructive agencies, the removal of dead material may assist in checking the ravages. This was clearly proven by the splendid results attending the bark burning campaign carried on by the Government in its warfare on the Black Hills beetle. Such a cutting, of course, is largely for salvage, and the question of keeping the canopy intact cannot be considered. Large holes are often made in the canopy which nothing but underplanting will repair.

Pruning.

When the trees in a stand were too widely spaced in youth to have pruned themselves naturally, artificial pruning is sometimes advisable. However, it is an extremely intensive procedure, and, for the most part, should be classed as an æsthetic, rather than an economic, measure. If young reproduction is interfered with, a careful pruning may produce excellent results at a trifling cost in comparison with planting.

Artificial pruning in the United States is largely confined to the removal of dead branches for a distance equal to one log length up the tree. Schenck's plan of cutting off green branches, leaving a stub six to eight inches long, and two or three years later knocking these stubs off with the poll of the axe has some points in its favor. Ordinarily, pruning of green branches is not favorably regarded, since the open wound permits the entrance of fungi, and in addition it is claimed that it causes loose knots.

White pine in pure stands often fails to clean itself properly. Its value as clear lumber is much higher than when knotty. In such circumstances the expenditure of $2.00 to $3.00 per acre in pruning one hundred of the best trees would be a sound investment that would ultimately yield splendid returns.

The regulation pruning tools are used, and for good-sized limbs a long-handled saw will prove most effective. In every case a clean cut should be left that will soon heal over.

Pruning in the United States can be done for two to three cents per tree, where only the lower limbs are taken off. In Germany, where all dead limbs, and green branches up to three inches in diameter are sometimes removed, pruning costs about six cents per stem. The per acre cost is kept down by confining the operation to those trees that will be harvested last, otherwise the cost would be prohibitive.

FIG. 18.—A CLEANING IN A MASSACHUSETTS WOODLOT—GREY BIRCH AND PITCH PINE WERE REMOVED AND REMAINING WHITE PINES HAVE BEEN PRUNED.

While cleanings may seem extremely intensive, it is often good management, since it costs $10 per acre to plant the seedlings, to spend $2 per acre in preventing the death of some thrifty young seedlings when half grown.

Economic Considerations.

While there is some objection to thinnings and improvement work, as unnecessary and unnatural measures, nevertheless, continental experience has proven that they are well worth while. It is true that the removal of trees from an immature stand causes a certain amount of damage to the remaining trees, but that is trivial in comparison with the retarded growth and misshapen boles that would result were thinnings not made.

In shallow-rooted species, thinnings may be impossible, but by beginning early and cutting judiciously, it is nearly always possible to remove some of the least desirable specimens.

Therefore, the question is not whether thinnings, cleanings, etc., are advisable, but whether the desires and financial status of the owner will warrant such an expenditure, and whether the increased value of the product will pay for the present investment.

The general statement may be made regarding American forestry conditions, that under ordinary circumstances improvement work should be postponed until the material is at least large enough to pay the expenses of removal. If the circumstances of the owner, and market conditions warrant a slight outlay, it should be made for the benefit of the future yield.

Any system of forest management necessitates a certain amount of sacrifice, even if profits are simply deferred, but to convince thinking men, the future returns should be assured.

The cost of improvement work, on the whole, is not excessive, and while thinnings for instance, may increase the cost of cordwood from ten to twenty per cent, over extensive methods, the investment on the whole, is worth making.

BIBLIOGRAPHY.

COOK, H. O. Improvement Thinnings in Massachusetts Woodlands. Published by Massachusetts State Forester.

GRAVES, HENRY S. Principles of Handling Woodlands. John Wiley & Sons, New York.

HAWLEY, R. C. Treatment of Hardwood Lands in Southwestern Connecticut. Forestry Quarterly, Volume V, page 283.

HAWLEY AND HAWES. Forestry in New England. John Wiley and Sons, New York City.

NISBET, JOHN. Studies in Forestry.

SCHLICH, SIR WILLIAM. Manual of Silviculture. Volume II.

CHAPTER VI.

ARTIFICIAL REGENERATION.

General.

In certain circumstances it may be necessary to start forests artificially, and where seed trees of the desired species are absent or it is desired to get a forest of a given mixture started at once, planting may be resorted to with good results.

The first mention of forest planting is made early in the thirteenth century when the cities of Nuremburg and Frankfort, Germany, attempted artificial regeneration. These results were more or less experimental and it was not until the beginning of the eighteenth century that forest planting was practiced on any large scale. Wildlings were used at first, which in turn gave way to artificially propagated seedlings, and in 1840 the use of transplants became quite general in German forest practice.

In the early history of the forestry movement in the United States, forest planting occupied quite a prominent position, but since the chief reason for most of the planting was to acquire land under the Timber Culture Act, short-lived species which made a quick showing were largely used, so that these plantations did not long endure.

From one point of view, extensive reforestation in the United States seems unnecessary, when half-grown timber may be bought in different parts of the country at a price less than the cost of planting. But, in view of the vanishing timber supply, the idle land, and water storage problems, reforestation on a large scale appears to be the only solution.

The planting of timber trees has great possibilities in the prairie states, where fence posts, building material, etc., are valuable, and where the indirect influence of windbreaks is also of great importance. In regions where large areas of second quality land are to be found in close proximity to good markets, forest planting promises to yield good results. For instance, there are thousands of acres in Massachusetts,

FIG. 19.—SITE SELECTED FOR EXPERIMENT IN FIELD SOWING—LOLO NATIONAL FOREST, MONTANA.

It is planned to reforest 20,000 acres annually for the next 10 years by planting seedlings and direct seeding together wherever the latter method is practicable.

Connecticut and New York, which could be planted with forest trees to good advantage, and raw material could be produced close to desirable markets.

The advantages of reforestation can be briefly expressed as follows:

1. As an economic measure, forest trees should be planted only on land too poor for agricultural purposes. This will bring land into use that would otherwise lie idle.

2. Planting will tend to raise the quality of timber that is produced. Our forest lands have suffered in the past from neglect, and have produced weeds rather than desirable timber trees. With more intensive management, much better material will be produced.

3. The experience of European foresters indicates that with the rise in stumpage that will inevitably occur, forest planting must yield a higher return than we anticipate now, and aside from making idle land yield, forest planting is really a splendid financial investment.

There are certain drawbacks, however, to the situation. Fire, of course, is apt to wipe out a plantation in a short time, fungus diseases, drought, etc., may also render such an investment precarious. Finally, excessive taxation might seriously disturb the estimated returns. The latter objection, however, is rapidly disappearing before the enlightened legislation which the different states are now enacting. None of the drawbacks are serious enough to deter any owner of idle land.

Status of Forest Planting.

Abroad. — Practically all the foreign nations are planting at the present time. France in controlling her mountain torrents has restored more than a million and one-half acres to forest cover. Switzerland, up to the beginning of the twentieth century had reforested over 16,000 acres. Russia, with its vast areas of virgin forests has planted extensively, and in Germany, the Prussian foresters alone are planting on an average of 50,000 acres a year. In the Karst region on the shores of the Adriatic, Austria-Hungary is planting to restore forest cover to an area laid desolate by excessive cuttings and fire.

In the United States. — Planting was started in the United States in 1893 and is now being extended as fast as material and funds will permit. At the present time extensive experiments are being carried on along lines of direct seeding. Small nurseries are connected with many of the ranger stations in addition to the large nurseries operated by the Forest Service, and the use of seedlings and transplants is on the increase. Besides the actual planting operations, the germination of seeds and the heredity of the various species have been tested, and the success of various methods of direct seeding has been proven.

Of all the states engaged in reforestation, New York takes the lead, closely followed by Massachusetts. New York at the present time has nine large nurseries with an output of about eight million trees per annum. In Massachusetts, the unique system of planting land loaned to the state for a period of years has been put into effect. The original owner can reacquire this land by paying the cost of operation within ten years. Various railroads, water companies, and paper companies are practicing reforestation on an economic basis, since they realize that any corporation permanent in its organization and owning large areas of land can practice forestry effectively and profitably, and, at the same time, provide for a future supply of raw material.

Private planting has always been practiced in the United States by individuals, with great success, and large plantations up to 300 to 400 acres are often established by land owners interested in forestry. Individual enterprise for forestry planting is apt to be checked by the thought of the length of time to elapse before harvesting the crop, but with a corporation or state, this argument has no weight. However, in certain sections of the country, individuals are planting forests with an idea of raising material on a short rotation that will beautify and enhance the value of their land, and at the same time yield a good income on the investment.

Seed Collection and Storage.

The seed is the resting stage of plant development, and is produced after several seasons of favorable climatic conditions. Some seed may be produced every year, but the abundant production occurs at intervals. Most seeds take but one year to mature, if proper fertilization is given, but black oaks and the pines take two years, and the junipers may take one, two, or three.

Most seeds ripen in the fall, but willows, poplars, elms and soft maples ripen in the early spring, and must be gathered at once. In nature, the seeds are disseminated in a variety of ways. Some are provided with wings to aid their transportation by the air, others are contained in a fleck of down, which enables them to be carried long distances. Some are berry-like in form and depend on the birds for distribution, while still others depend on gravity.

The time to gather seed depends on the time the seed or the fruit matures. It should be gathered as soon as it is ripe, but before nature starts to scatter it. If seeds are wind-sown, they should preferably be picked from the tree, but sometimes such seed as elm, maple and ash may be picked from the ground in large quantities. With conifers the cones are either pulled off with rakes, or if a lumbering operation is available, the cones may be picked from the fallen trees.

Trees in middle life produce the best seed, also those that have had good light in youth.

Cost of Collection. — The cost of seed collection varies greatly with the species, the season, and the methods used.

FIG. 20. — TRANSPLANT BEDS, GOVERNMENT NURSERY, HELENA NATIONAL FOREST, MONTANA.

There are now 17 large and 21 small nurseries operated by the Forest Service having an annual output of over 20,000,000 seedlings.

Spruce seed during good seasons may be gathered on a logging or lumbering job at a cost of 30 cents a bushel for the cones, or 47½ cents a pound for the seed itself. From this figure, the price may range up to a dollar or more a pound for collections under more adverse conditions. The locality in which seeds are gathered has considerable bearing on the future seedlings, as seedlings collected in a southern climate are apt to prove very weak if set out in a colder region. Black walnuts gathered in Missouri, for instance, were killed by spring frosts in Minnesota, when those grown from local seed came through the season unscathed.

Storage of Seed. — While nature plants seeds practically as soon as they are mature, it is a better plan to keep the seed or fruit over the winter and plant it when the weather conditions are more advantageous. The seed of species like elm, maple, poplar and willow are hard to keep, however, and ordinarily must be planted at once. Seeds that usually keep over winter must be stored dry, while those that dry out must be given additional moisture, and enclosed in air-tight receptacles. Most fleshy fruits require stratification, which consists of laying alternate layers of fruit or seed and moist sand in a box and burying it on a hillside, where drainage is good. Small seed like ash, for instance, which tends to " lie over " unless forced, could be stored between layers of cheesecloth covered with sand, and when planted will ordinarily germinate immediately. Commercial houses rarely practice stratification, because they have storehouses where moisture conditions are kept uniform. Nuts and acorns may be kept in bins, while small seeds can be kept to advantage in large glass bottles or carboys. Whatever form of storage is used, care should be taken that rodents can not reach the seed.

Seed Extraction. — At the present time, the extraction of seed from cones is carried on with considerable skill by seed companies in the United States, and the Federal Forest Service has recently commenced to gather and extract coniferous seed on a considerable scale. In the National Forests, cones are often gathered in inaccessible places and are carried to an extracting station in large bags or canvas cloths. They are spread out on canvas in thin layers and exposed to the full sunlight to hasten drying. From time to time, the cones are stirred, and are covered during the damp weather and at night. As the cone scales open under the drying influence of the sun, the seed falls out upon the canvas sheets. A large amount of the seed falls out naturally in this way, but as a final resort, the open cones are put in a churn, made from a rectangular box with slatted sides, and turned until all the seed is threshed out. The seed is then cleaned by crushing the wings against a sieve, and after the chaff has been removed by winnowing, it is packed in small bags for transportation.

The method adopted in commercial seed-extracting establishments is the same as the above in principle. The cones at such plants, however, are ordinarily allowed to dry outside for a few days before exposing them to artificial heat. In the drying room the temperature of 120 degrees Fahrenheit will cause the cones to open in a few hours, and from this on it is merely a question of threshing the seed and removing the wings and chaff.

German seed-extracting establishments lay great stress upon the fact that the seed must not be injured during the process, as if the seed coats are cracked in any way, fungus spores can find entrance, or the cotyledon will be acted upon first in germination, which will make the seedling practically useless for planting. While extracted seed can be kept over until the next season with little loss, cones cannot be so retained, as the seed will rapidly deteriorate owing to the moisture present within the cone.

Commercial seed houses ordinarily test their seed from time to time, and the results are carefully recorded for each batch of cones. This serves as a check on the quality of the cones and on the technique of the extraction. Seeds are tested by being placed on plots of sand or moist blotting paper and exposing them to uniform heat and moisture conditions.

Direct Seeding.

There are two general methods of planting seed directly in the field: broadcast, and partial seeding.

In *broadcast* seeding all parts of the area receive the same amount of seed; but for this method to be successful, the soil must not only be good, but it must have the faculty of retaining moisture in the superficial layers long enough to induce germination. If the downward growth of the root can keep ahead of the drying out of the soil, broadcast seeding will ordinarily result successfully. Some preparation of the soil is necessary, as a rule, since it is practically useless to sow seeds broadcast on a thick sod or on ground densely covered with weeds. In open woods with loose soil, broadcast seeding may be advantageous if seed can be cheaply procured.

The Federal Government has had some success in the Black Hills with broadcast seeding on burned-over lands, but there it has been found, however, that aside from conditions of soil and season, the presence of rodents has a very important bearing upon the outcome. From the experience gained in that region, it is believed that squirrels, field mice, etc., must be poisoned off before broadcast seeding is attempted, otherwise it will prove a failure.

To get a thick stand of seedlings in eight or ten years, a large amount of seed is required, and unless very cheap seed is available, planting is apt to be more economical and certain than broadcast seeding. In broadcast seeding experiments conducted in North Carolina, the following amount of seed per acre was found necessary: Ash 40 pounds, maple 40 pounds, white pine 12 pounds, spruce and larch 10 pounds, yellow pine 8 pounds. Even at a cost of 50 cents a pound for seed, it will be seen that broadcast seeding is doubtful economy, and, as a rule, there will be a closer and more uniform stand if seedlings are used.

In *partial* seeding, the seed is not uniformly scattered, and, on the whole, is much better than broadcast seeding, on account of the decreased cost. Partial seeding has some advantages over the preceding method. Less seed is used and, therefore, only the best places are selected for seeding, and the balance left untouched. This means better covering of the seed, with greater chances of success. Partial seeding may be practiced either as strip seeding or spot seeding In the strip method, the seeds are sown on strips of cultivated or broken soil, with untouched rows alternating. If weeds or sod are present, the strip of cultivated soil varies in width depending upon the intensiveness of the operation and the character of the ground. In ordinary circumstances, the cultivated strip will be one-eighth of the untouched portion.

This method has been used somewhat in the past and several cases are on record in New England where pine cones were planted in furrows, giving a decidedly dense stand.

Seeding in spots is the most common method of direct seeding employed. The seed spots vary in size from four inches square to thirty square feet, depending on the size of the seed sown, on the condition of the soil, and on the kind of ground cover. Direct seeding of this kind can be used in places where a plow cannot run, and where the woods are open only in spots. The ordinary practice is to scratch the surface of the soil for about three inches square, and if spaced six feet apart, there will be 1210 seed spots to the acre. When sowing on heavy sod, it is customary to remove it and after

FIG. 21. — FOREST OFFICERS PLANTING TREES ON A BURNED-OVER WATERSHED, PIKE NATIONAL FOREST, COLORADO.

A planting crew of 2 men can plant from 800 to 1200 seedlings per day, depending on the soil, ground cover, and slope.

working up the soil, place the seed on the loose dirt, and cover it lightly. Field planting by means of the spot method in Massachusetts and New Hampshire has given fair results, although the late spring frosts are apt to heave the young seedlings to a marked degree. Under the most favorable conditions, seed-spot methods may work to advantage, but outside of the open woods they are protected from frost with difficulty.

With tap-rooted hardwoods, however, the planting of the seed directly in the fields is desirable, and results are apt to be better than when the seedlings are used. In this case also, rodents or birds are apt to do considerable damage, so that the acorns and nuts should be treated with a heavy coat of red lead or other poisonous preparations before planting.

Results and Costs. — The stand of seedlings obtained by direct seeding may vary widely, depending on soil, season, presence or absence of rodents, and method of distributing seed. As stated above, unless seed is extremely cheap, or the use of seedlings made impossible by the size or the accessibility of the area to be planted, direct seeding is apt to be less satisfactory than where stocky seedlings are planted. The costs of direct seeding operations, labor and seed taken together, are as follows*:

Method.	Species and pounds per acre.		Cost per acre.
Broadcast	White pine	10 pounds	$15.00
Broadcast	Yellow pine	8 "	5.37
Broadcast	Austrian pine	4 "	4.00
Simple spots	White pine	1 "	3.00
Simple spots	Yellow pine	1½ "	3.92
Simple spots	Austrian pine	1 "	3.96
Simple spots	Douglas fir	1 "	10.84
Raked spots	Austrian pine	1½ "	2.28
Harrowed	Austrian pine	5 "	2.63
Corn planter	Austrian pine	1 "	1.49
Corn planter	Yellow pine	4 "	4.63
Carefully prepared seed spots	Austrian pine	2¼ "	31.66

Sod removed one foot square and earth carefully worked four to six inches deep, and seed then covered lightly. Yellow pine refers to western yellow pine (Pinus ponderosa).

* J. Murdock, Jr., Forestry Quarterly.

From the above figures, it will be seen that the range of prices is great, depending on the intensity of the operation, as well as quantity and cost of the seed used. In the above instances, the Austrian pine seed cost but 35 cents per pound, this accounting for the comparatively low cost per acre.

Nursery Practice.

By all odds, the most widely used method of reforestation is that of planting nursery seedlings or transplants on the area where forest cover is desired. This method, while apparently more expensive, has proven, for the most part, to be the most economical in the end.

Seed and Transplant Beds. — No point in nursery practice is more important than the selection of a desirable site for the forest nursery. Not only are good soil and plenty of water necessary, but drainage and slope, as well as accessibility, should also be kept in mind.

The soil should be a mellow loam, free from stones, well drained and free from weed seeds. The ground should be cultivated a year previous in order to free the soil from any weeds, and any attempt to raise seedlings on a situation which was in sod the year previous is apt to result in a heavy loss, owing to the action of the larvæ of the May beetle on the roots of the young seedlings.

With the site of the nursery selected, the size will depend on the scope of the planting operation; one-half acre will produce 145,000 three-year transplants annually.

If land is available, it is wise to have the nursery large enough to permit rotation of crops, so that seedlings, grass, and transplant beds may alternate on the same area. The beds should be worked up in the spring as soon as the frost is out of the ground, and laid out running east and west preferably, in order that proper sunlight may be available. Four feet is the standard width for beds, and the length may vary according to the plot. Four by twelve feet is taken as a standard, although modern nurseries have beds up to one hundred feet long. Heavy fertilization is the custom, and two barrels of rotted leaves and the same amount of well-rotted manure to fifty square feet of ground is the ordinary application. The ground is well mixed with the manure and the mixture should be finely pulverized, before the seed is sown.

FIG. 22.—PLANTATION OF WHITE (LEFT) AND SCOTCH PINE (RIGHT) AMANA, IOWA.

Artificial pruning has been practised on both species about 10 feet up the bole.

The beds are ordinarily raised slightly above the ground in order to insure good drainage, and the surface of the bed should be free from pockets, so that water will not collect in the spring. Aside from raising the beds three or four inches above the level of the path, it is customary to slightly crown the surface of the bed in order to secure better drainage.

Side braces or frames are largely a matter of convenience.

The most intensive nursery practice calls for well-built frames with netting surrounding each bed, whereas many commercial nurserymen merely sink a three-inch board about one inch into the ground. In small nurseries, wire screens made of three-fourths inch mesh are used to cover the beds, so that the birds will not tear the bed to pieces by picking at the husks of the seeds.

Screens of lath, placed so as to make half shade, are used to prevent too much light. These are best used in lengths of six feet, as one man can handle this size. High screens, made of lath or brush, are used sometimes, but with high shade advantage cannot be taken of any dull day to harden the seedlings, and, on the whole, low shade maintained by the use of lath screens fitting directly over the frame is considered more desirable.

After the beds and screens have been prepared, the sowing can be started. The amount of seed used for one bed, four by twelve feet, is approximately as follows *:

Species.	Ounces per 4'×12' bed.	No. of seed per pound.	Average germination, per cent.
White pine	10	30,000	60–70
Red pine	6	80,000	70–80
Scotch pine	8	70,000	60–80
Pitch pine	10	50,000	65–85
Jack pine	6	90,000	60–80
Norway spruce	8	65,000	60–70
Red spruce	6	120,000	65–80
White spruce	8	100,000	60–70
European larch	16	75,000	50–60
Balsam	12	50,000	35–60
Arborvitæ	6	175,000	60–70
Hemlock	8	80,000	35–65

The seed bed is generally moistened thoroughly before sowing is started, and the seed is then lightly covered by

* Forest Service Bulletin 76, C. R. Pettis.

passing loam through a fine sieve, and then the surface of the bed is firmed with pressing with a clean hoe or plank.

One of the most important points to be observed in nursery planting is to secure absolutely sterile dirt to cover the seed. "Damping off," a fungous disease attacking the young seedling during the first few weeks of its existence, is especially apt to be present when the loam of garden soil has been used. Consequently, sterile soil secured from three to four feet below the surface should be used in covering the seed. The beds are then covered with the screens, and burlap or leaves may be placed over the surface of the bed to keep the ground dark and moist until germination is completed.

In ordinary circumstances germination will take place in fourteen to sixteen days for the conifers. Care should be taken not to delay the removal of the burlap or leaf covering, otherwise the young seedlings may be entirely smothered. The lath shade is kept on continuously during the first summer, except during cloudy days, when it may be removed to give more light to the seedlings and to prevent the danger of "damping off." Drought is also to be guarded against during the first season, and any prolonged period of dry weather should be supplemented by careful sprinkling. Weeding is particularly heavy the first season, especially if the land was not well cultivated previously. Weeding and attention during the first year costs from 9 cents to 20 cents a thousand.

At the end of the first year, the cost of seedlings raised on a scale of two million will be from 55 cents to 65 cents per thousand in the bed; but overhead charges on a small nursery would raise these figures. Weeding is ordinarily discontinued about September first, and whatever grass or weeds grow afterward are allowed to remain, in order to protect the roots of the young seedlings during the winter. The lath screen is also removed, late in the fall, to harden the seedling, and, as winter sets in, a mulch of weeds should be thrown on top of the seed bed or a strip of burlap placed on top of the first fall of snow to prevent rapid thawing out during the following spring. If the operation has been carefully carried out, there should be approximately 200 seedlings per square foot, or 9000 to 10,000 seedlings for the unit bed.

In the spring of the second season, the bed can be moistened and seedlings can be pricked out, leaving from 75 to 100 to each square foot. Those that remain will have additional room for their development and those removed can be put in the transplant row.

At the beginning of the second season, the bed should be uncovered as soon as the frost is out of the ground, and careful watch should be kept to avoid heating by retaining the cover too long. Attention during the second season is much lighter, consisting of a little weeding and an occasional watering. The tops, by this time, will form a dense mat which will keep the ground moist and will crowd out weeds. At the end of the second season no mulch is necessary as the roots are now deep enough to prevent heaving.

Ordinarily, the seedlings do not remain longer than two years in the seed bed, and at the beginning of the third season, the plants are generally removed and placed in transplant rows for better development of their root system. The cost, at the end of the second season, is approximately $1.85 per thousand in the bed. If shipping is contemplated, thirty to forty cents per thousand for digging and packing should be added to this cost.

Planting material should not be too toppy, as the main advantage in artificially grown seedlings is the character of the root system rather than the size of the tops. Too large material is a drawback, being bulky and not able to stand the shock of removal; consequently, a year or two in the transplant beds is very beneficial. These beds consist of rows twelve to fifteen inches apart with the seedlings placed from one and one-half to three inches apart in the row. This spacing will permit the planting of from 150,000 to 300,000 seedlings per acre, and by the use of long beds, weeding with machines can be practiced, which will greatly reduce the cost. In some parts of the country, the seedlings are allowed to remain but one year in the seed bed and then two years in the transplant bed. This "1–2" seedling, so-called, is quite desirable for certain situations. The more common way is to permit the seedling to remain two years in the seed bed and either one or two years in the transplant bed, making 2–1 or 2–2 transplants. The transplanting in rows was first done by hand, but the Yale transplanting board, invented by Prof. Toumey of the Yale Forest School, has simplified this operation, increased the speed remarkably, and greatly reduced the cost.

Field Planting.

Reforestation of open country is by far the most common form of forest planting, although underplanting is being practiced to some extent. The idea of establishing forest covers on watersheds, and providing for a future supply of timber, are the chief economic reasons for planting, although the æsthetic value of forests should not be overlooked. Shelterbelts, however, are extremely important in certain parts of the country. Investigators claim that, in the states of Kansas and Nebraska, where enormous losses are suffered by the farmers owing to the hot winds which sweep up from the south, twenty per cent of the area could be planted to shelterbelts running east and west, and the remaining eighty per cent of the land would produce as much in agricultural crops as the total does now. In addition the comfort of the inhabitants would be vastly increased, and there would be considerable incidental revenue derived from the timber production of these forest-covered strips.

Choice of Stock. — The following points to a large degree control the kind of planting stock used:

1. *Soil and Moisture.* — The kind of soil and amount of moisture practically control the choice of species. Hardwoods must have fairly rich and moist situations as a rule. Among the conifers, white pine may be used on medium sandy loam, but Norway, pitch and western yellow pine, for example, may be used on the lightest and sandiest soil.

2. *Climate.* — The bearing of climate is very important also, for it is almost useless to plant a species outside of its geographic range, or in a region of decidedly different climate. According to Mayr, zones of like climate are to be found in different parts of the world, so that Japanese species may be planted successfully for ornamental purposes in the eastern part of the United States. On the whole, it is not practicable to plant western species in the East, or northern species in the South.

3. *The Desire of the Owner.* — The ultimate plan of the owner should be thoroughly understood, and a sharp line should always be drawn between æsthetic and economic planting.

4. *Density of Shade and Amount of Ground Covered.* — Only tolerant species could be used where there is a dense growth of weeds.

5. *Market Demands, Present and Future.* — If a plantation is to be made from an economic standpoint, the future as well as the present market possibilities should be taken into consideration. In Massachusetts, the white pine is now and probably always will be the most saleable species, owing to the box-board industry. In western New York State and in the middle west, fast growing, durable species, like black oak,

chestnut, or catalpa could be used to produce vineyard stakes, fence posts, etc.

6. *Resistance to Attacks.* — Species should be chosen with a special reference to the freedom from diseases which may be prevalent in the region. Oaks are particularly subject to attack by the gypsy and the brown-tailed moths in Massachusetts. The chestnut-bark disease may make the planting of chestnut very hazardous throughout its whole range, and white pine, in certain portions of the East, suffers severely from the ravages of the white-pine weevil. As a general recommendation, it is not safe to decide on the species to be used without study of the local conditions. The soil and moisture conditions should be noted on the ground, and a careful study of the market situations and the desire of the owner made before choosing species.

Age. — Nursery stock is classified as seedlings, either one, two, or three years old, generally sold on a basis of size, or transplants which have been put in beds one or more times to improve the root system. Complete specifications of a transplant cover the number of years in the seed bed, the number of years in the transplant bed, number of times transplanted and the size. For instance, a 2–2 transplant, once transplanted, 14 inches tall, would be the nurseryman's method of describing stock. The price, of course, is increased by the extra handling, and, as a rule, a seedling the same size as a transplant is worth about half as much. As a general statement, it may be said that the smaller the planting material used, the more successful will be the operation, since the younger stock can stand the shock of field planting much better, can be handled much more easily, and costs considerably less.

Transplants, of course, can stand more drought and shade, and must be used where these conditions are met. Most broadleaf species are put out as one year stock, while conifers will average from two to four years of age when finally planted. Scotch pine, however, is sometimes planted at one year, but one-year coniferous stock is rarely planted successfully in the United States.

Technique of Planting. — On the arrival of the material on the planting site, the hampers or baskets containing the seedlings are opened and the bundles which are usually packed in moist sphagnum moss are taken out, and loosened, the roots dipped in a puddle of mud, and then the loose bundle is laid on the side of a slooping trench, the earth thrown upon the roots and gently firmed. This is done to prevent the fine root hairs from being dried out. Plants can be kept for three weeks if necessary, provided the situation is shady. The tools needed in a planting operation are mattocks or spades and water-tight pails. The best results are obtained by keeping a layer of soft mud two to three inches thick in the bottom of the pail in which the roots are kept moist. The planting crew consists of two men as a unit, one using the mattock, the other inserting the tree in the hole made by the mattock man. Ordinarily, these men can keep fairly accurate lines, but if especially straight lines are desired, it may be wise to stake the field off or use lines stretched across the area.

The ordinary planting space is six feet apart each way, for economic planting, which requires 1210 trees to the acre. A planting crew can cover from two-thirds to one and one-fourth acres a day, depending on the size of the plants and the character of the soil.

Shelterbelts. — In certain parts of the West, where hot south winds are frequent, the use of windbreaks is strongly urged. Aside from greatly increasing the comfort of the owner, it has been proven that the beneficial effects of windbreaks upon the growing crops is most marked, and, by correct placement they will insure the proper distribution of snow and prevent disastrous piling. The force of the winds can be greatly checked by planting windbreaks, running east and west, at a distance of one-quarter of a mile, or less, apart. The gardens or orchards in the lee of the breaks will be considerably benefitted, the comfort of the family will be greatly increased, and the incidental returns in the form of fuel, fencing, etc., will more than repay the cost. The species giving best results for windbreak planting should be quick growing, hardy and able to form a dense break. It is extremely difficult to find one tree that possesses all of these qualities, so mixtures are often resorted to.

The poplars, either Carolina or Norway, the box elder and silver maple are fast growing trees; when planted close, they serve well for a few years, on account of their rapid growth. To achieve the best results, they should be reinforced by some hardier tree like the white elm, Russian wild olive, green ash, or white willow. The Russian olive will grow remarkably well in poor alkaline soils, but it should be replaced by some of the others in better soils.

Evergreens, of course, make the best windbreaks, but their growth at first is decidedly slow, and a nurse tree is often necessary. The Norway spruce, white pine, white cedar, the western yellow and jack pines, are the most favorable species for this use, although the spruce requires rather moist soil as a rule.

The planting distance will range from four feet to eight feet, depending on the growth rate and the branching habit of the tree. Rather close planting is desirable as the effect is felt at once, and as soon as crowding is noticed, alternate trees may be cut out. The width of the windbreak is also of importance. Trees in a belt as in the forest can protect the ground and assist each other far better than two or three rows of trees. The breaks rarely exceed two rods in width, as the snow is apt to pile in and break down the trees. In some cases, two narrow belts of trees may be planted and a small garden planted between them.

Contrary to regular forest planting, cultivation during the first few years is practically indispensable, and under certain conditions, mulching is quite advantageous. After rapid growth has started and the ground is well shaded, the trees can take care of themselves.

Planting Under Cover. — Throughout what might be termed the entire sprout hardwood region, countless woodlots may be found which have been abused and neglected for many years. The desirable species have been removed for lumber, and the ground left practically at the mercy of the forest weeds. Fires have been permitted to run without hindrance, and cattle have been allowed to graze at will, destroying all chances of natural regeneration. As a consequence, we find thousands of acres, where the mature trees are going to pieces, and such reproduction as is present is confined to undesirable species. In such cases, planting under cover is an admirable solution of the problem of building up the woodlot. Conifers are favored, for the most part, because the forests described are largely composed of sprouting hardwoods, and a mixed forest is a desirable goal at which to aim. However, hardwoods of rapid growth and high technical value like ash, basswood, red oak, and tulip poplar may be inserted to advantage, but, for the most part, spruce and pine as three- or four-year-old transplants are the most desirable stock to use.

The planting operation is practically the same as in open field planting, and the makeup of the crew the same; only instead of lining up on a stake or landmark, the crew should endeavor to pick out likely spots in the stand. Beneath a hole in the canopy or where there is a moist bit of soil, or where there is any open space on the forest floor is a good place to insert the seedlings. Tolerant species like spruce must be used if the canopy is comparatively dense, but the transplants can be used on lighter soil, provided there is sufficient light coming through the cover. Such an operation permits the owner to thin out his woodlot later on, and allows a choice between hardwoods and softwoods with the idea that a mixed forest can be obtained. It is needless to say that after an operation of this sort, the woodlot can no longer be used for grazing.

Spacing and Costs. — The ordinary spacing for open planting is six feet by six feet, which means 1210 trees to the acre. Where trees are to be planted for a short rotation, as willow for basket withes, black locust or catalpa for fence posts, or poplar for excelsior, closer spacing may be made. In Germany, where planting material is cheap and land is valuable, the foresters plant intensively, sometimes using as high as 5000 seedlings per acre. The following figures will give an idea of the number of trees per acre at the different spaces:

Spacing.	No. of trees per acre.	Spacing.	No. of trees per acre.
4×4	2722	7×7	888
4×5	2178	8×8	680
5×5	1742	9×9	542
5×6	1452	10×10	435
6×6	1210	....	....

In under-planting, the number of trees to be used depends largely on the condition of the forest canopy and the desire of the owner. When the canopy is comparatively dense, with but few holes in it, it is useless to plant more than 100 to 200 per acre, unless a thinning is made previously. If, however, the stand is quite open through heavy thinnings or repeated fires, 500 to 600 seedlings may be put in to advantage. In under-planting, either in the average woodlot or in young reproduction, it is possible to take advantage of the sprouts and seedlings as side shade, and place the transplants five to six feet from an oak or maple sprout, and thus reduce the amount of planting material without impairing the result.

The cost of a planting operation varies widely with the density of planting, the size of material used, and the character of the soil. If large stock is used, to make an immediate showing, the price may be prohibitive to the average owner. By using the smallest planting material possible and allowing for no preparation of the soil, the average cost per acre for open planting, six feet apart, ranges from $7.00 to $10.00 an acre. For under-planting, where stockier transplants must be used to withstand shade and root completion, the cost of planting material is correspondingly higher, but the fewer seedlings used reduces the price. Where transplants can be obtained at a cost not exceeding $6.00 per thousand, a rough estimate of one cent per tree in the ground is safe; that is, under-planting with 400 trees per acre will cost approximately $4.00.

Yields. — Reforestation is being urged by foresters chiefly as an economic proposition, although from an æsthetic point of view, planting has its advantages. Unless it can be definitely proven, however, that the money invested will yield a fair return, forest planting had better be abandoned, or else charged up to landscape improvement. While American forest practice is not old enough to possess the wealth of experience which is available in Germany and France, data are, nevertheless, at hand to prove conclusively that plantations can yield a fair financial return on the capital invested. Forests, however, are a long-time investment, and with fair fire protection are quite stable, so that the required interest rate should not be put too high. Where estate owners have idle land, or in the case of paper concerns desiring continuous supplies of pulp wood, etc., reforestation is a highly desirable investment, and one worth the attention of shrewd financiers. Already, railroads and paper companies in the East have begun extensive reforestation to provide them with future supplies of ties and pulp wood. Water companies whose drainage areas cannot be grazed or fertilized, for fear of contaminating the supply, are finding forestry an ideal solution of their land problem. Practically any corporation owning land and with an assured continuous existence can plant trees for future timber supply with profit.

As intimated above, on account of the stability of a forest plantation as an investment, and because of its duration, thereby avoiding frequent reinvestments, forests should rank as bonds. Therefore, a comparatively low rate, say 3 per cent to 3½ per cent is all that should be expected. However, investigations have shown that if a plantation of white pine can be established at a cost of not more than $15.00 per acre for land and planting costs, 5 per cent compound interest is the assured return on the investment. This yield, moreover, is based on present stumpage value, and the certain rise in the price of timber products will cause a remarkable appreciation in this rate. The following table shows what may be expected from a stand of white pine planted on cheap land at an average cost.

The final column in the table shows the net profit over and above the five per cent rate. For instance, at fifty years, based on present stumpage values, such a plantation would yield $214.10 more than a five per cent investment.

The most frequent criticism that is levelled against forest planting by the individual is that the investor rarely if ever

FINANCIAL ROTATION OF WHITE PINE.*

Money valued at 5 per cent; value of land, $4 per acre; cost of planting, $10 per acre.

Age of stand (years).	Gross returns (stumpage value).	Expenses and interest.									Gross profit.	Net profit over 5 per cent on investment.
		Taxes.					Cost of producing.		Total investment without interest.	Total investment and interest at 5 per cent.		
		On timber.			On land.							
		Annual for five-year periods	Amount paid in taxes.	With accrued interest.	Amount paid.	With accrued interest.	Interest on value of land.	Cost of planting and accrued interest.				
25	$40.50	$0.540	....	....	$1.50	$2.50	$9.54	$33.86	$11.50	$45.90	$29.00	—$5.40
30	75.00	1.000	$2.70	$3.13	1.80	3.48	13.30	43.22	14.50	63.13	60.50	11.87
35	195.20	1.269	7.70	9.80	2.10	4.73	18.06	55.16	19.80	87.75	175.40	107.45
40	262.40	3.498	14.04	19.85	2.40	6.54	24.16	70.40	26.40	120.95	230.00	141.45
45	324.80	4.330	35.53	45.60	2.70	8.37	31.94	89.35	44.23	174.86	280.57	149.94
50	465.00	6.200	53.18	83.36	3.00	10.99	41.87	114.67	66.18	250.90	398.82	214.10
55	505.50	6.740	84.18	142.37	3.30	14.30	54.54	146.35	97.48	357.56	408.00	148.00
60	532.00	7.080	117.88	220.82	3.60	18.55	71.70	186.78	121.48	497.85	410.52	34.15
65	566.00	7.548	153.28	331.40	3.90	23.97	91.35	238.38	163.28	685.10	402.72	—119.10

* From "Forest Mensuration of the White Pine" by Harold O. Cook, page 27

reaps the benefit. This, to a large extent, is true, but short rotations are possible where fence posts, Christmas trees, etc., are raised and in addition, the increasing negotiability of half-grown forests is a factor that must be reckoned with. If a man does not care to invest in a forest plantation for the benefit of his children or his estate, he may console himself with the thought that aside from the pleasure derived in watching the trees grow, the land covered with an embryonic forest will be much more saleable than stony fields or scrubby pasture.

Willow Culture. — The bulk of forest planting in the United States is done with the seedling or transplant as the stock. These require a considerable outlay in the form of nurseries, and the attention required to raise good nursery stock brings up the cost to a good figure.

Certain species can be propagated by simply planting a piece of young wood, preferably the last season's growth. Species like poplars and willows, are frequently propagated this way. The raising of willow for basket ware is quite an industry in certain parts of the country. It was introduced into New York by German immigrants between 1840 and 1850, and has since spread to other states. The site best suited for a "willow-holt" is a well-drained sandy loam containing a fair amount of humus. In some regions land subject to annual overflow is chosen, since this obviates fertilization. The American willow produces a tougher and more durable rod, but many European species are used, among which the Caspian and Welsh willows have found great favor.

The sets are planted either in the spring or late fall, after the growth has ceased, and are planted in rows fifteen or twenty inches apart, and the sets placed from five inches to nine inches apart in the row. This takes from 35,000 to 80,000 cuttings per acre.

Cultivation of the holt is practically unavoidable if good

growth is desired. If cultivation is difficult, the weeds may be mowed to prevent them from choking out the willows. After the stools are established, it is customary to cut the holt over annually. As soon as the leaves fall, the harvesting may commence and often continues through the winter. After cutting the rods are steamed, dried, sorted, and then packed for shipment.

The yield of willows ranges from four to seven tons per acre, with the price ranging from \$15 to \$25 per ton. With suitable land capable of being plowed for the prevention of insect and fungus attacks, willow culture is a profitable undertaking, and represents one phase of forest planting where returns are not long deferred.

BIBLIOGRAPHY.

Fetherolf, James M. Forest Planting on the Northern Prairies. U. S. Forest Service Circular 45.

Forestation in the Intermountain Region. Forestry Quarterly, Volume VII, page 127.

Forest Planting Leaflets, published by U. S. Forest Service.

Miller, Frank G. Forest Planting in Eastern Nebraska, U. S. Forest Service Circular 45.

Murdock, John, Jr. Results of Direct Seeding in the Black Hills, Forestry Quarterly, Volume X, page 27.

Piper, William B. Forest Planting in Northern Michigan. Proceedings Society of American Foresters, Volume VII, No. 2.

Pettis, C. R. How to Grow and Plant Conifers in the Northeastern States, U. S. Forest Service, Bulletin 76.

Report of Fifth National Conservation Congress Forestry Committee, entitled, "The Conditions under which Commercial Planting is Desirable."

Stephen, John W. The Basket Willow. Bulletin 3, New York Conservation Commission.

Waldron, C. B. Windbreaks and Hedges. Bulletin 88, North Dakota Experiment Station.

CHAPTER VII.

FOREST PROTECTION.

Forest protection is that part of forestry which deals with the protection of timberlands against destructive and injurious agencies such as fire, insects, fungi, wind, sand dunes and other minor influences, such as grazing, frost, sun scald, etc. The most important are taken up in some detail.

Forest Fires.

Of all the enemies of the forest, fire is the greatest. It is one of the most serious drawbacks to the practice of forestry in this country in that, under present conditions, it is considered a hazardous risk to hold standing timber. Remarkable progress, however, is being made, and at the present time about one-half of the remaining standing timber is receiving at least some attention in the matter of fire protection.

In placing any forest tract under scientific management, it is absolutely necessary to first insure safe protection from fires. A good portion of our early American forestry practice will be devoted for some time to securing efficient protection of our forests against the destructive waste of forest fires.

It is estimated that in the past the value of standing timber destroyed in this country by fire has been equal to that actually cut down and used. For the past forty years records of the amount of damage by fire show that an average of $50,000,000 worth of timber has been destroyed every year and that at least 20,000,000 acres of forest land or an area nearly four times the size of Massachusetts is burned over annually. To express the amount of damage in another way, if fires had been prevented on forest lands that are now unproductive on account of their destructive work, thirty-two billion board feet or four-fifths of the total lumber cut in the country could be produced continuously every year on this area.

FIG. 23. — A FOREST FIRE ON THE ANGELES NATIONAL FOREST IN CALIFORNIA.

Forest fires have burned up as much timber as has been used in the past. Both public and private organizations are adopting measures to prevent and control them.

During unusually dry years the losses from forest fires have been enormous, especially in those regions where the timber growth is dense and therefore favorable to the spread and disastrous effects of fires. For instance, in 1894, the famous Hinckley fire in Minnesota burned over an immense tract of timber, wiped out nine towns, six hundred people were killed and $25,000,000 worth of property was destroyed. In 1910 forest fires in northern Idaho and western Montana destroyed over $25,000,000 worth of white pine, yellow pine, Douglas fir and western larch timber in the short space of scarcely three weeks.

Kinds of Fires. — In general, there are two kinds of forest fires: ground and crown fires. The ground fire burns along in the grass, leaves or litter on the forest floor, or it may burn in the thick duff. In the last-named case it may burn very slowly and hold fire for many months, even burning as late as in mid-winter in the decayed accumulation of vegetative matter, such as leaves, twigs, etc. Ground fires of this class are sometimes separated as a distinct kind from the ordinary surface fire which burns along on the leaves and twigs above the vegetable mould.

Ground fires are especially common in the open grassy pine forests of the South and in our hardwood and coniferous forests in the North. They do considerable damage in the farm woodlots throughout the East. When fanned by a high wind and in the presence of sufficient inflammable material, such as brush, fallen trees and dense reproduction, a ground fire may develop into a crown fire.

The crown fire is the more serious kind and burns with great rapidity through the crowns of the trees often consuming everything in its path up to branches three inches in diameter. They occur only in dense coniferous stands or where the principal growth is of evergreen stock. Most conifers are highly resinous, especially in their bark and leaves, therefore they are very inflammable. Fires often run up the resinous bark of a tree or through dense underbrush from ground to crown fires. Some of these fires have been known to burn at the rate of six to ten miles an hour. Under the pressure of a

FIG. 24. — BRUSH BURNING IN WINTER ON A NATIONAL FOREST.

By eliminating the brush, the danger from forest fires is greatly decreased. Several states now require certain forms of brush disposal on private logging operations. It is commonly used on our National Forests in connection with timber sales.

high wind burning brands are often carried ahead of the main fire and ignite new fires in advance. It is practically impossible to stop the progress of a crown fire in a severe wind. Back firing is often resorted to but more often crown fires will become ground fires after passing the brow of a hill or ridge where it can be stopped by trenching.

The severity of any forest fire depends upon several factors. Perhaps the most important, aside from the character of tree growth, is the dryness of the forest itself. After a protracted drowth the fire hazard is very great and a small fire is likely to develop into a serious one, especially in the late summer. Some of our worst fires have occurred in September and October and have only been stopped by a heavy rainfall or because the continuous forest cover was broken by an open plain or agricultural land.

The character of the topography is also a very influential factor. Fires burn with great severity and rapidity up a steep slope, because the heat of the fire itself creates a draft and tends to intensify the progress of the fire. In burning down a slope a forest fire always progresses slowly and burns lightly over the ground so that it can be most successfully fought and controlled just over the brow of the ridge or at the foot of a slope.

The amount of inflammable material on the ground obviously influences the character and severity of the fire. The spruce woods of the Adirondacks are very brushy and dense and a fire, when once started, burns with great heat and rapidity. A fire in hardwood leaves is much less severe than in pine or spruce needles. The soil is also a factor in fire protection. The sandy plains of the east coast and the Lake States dry out with great rapidity on account of their porous nature, and ground fires are prevalent whenever adequate protection is not afforded. In Massachusetts pitch pine, a very fire-resistant tree, is often the only tree left after repeated fires have burned over the sandy plains. In portions of the Northwest the clay and loam soils keep the forest floor in a comparatively moist condition and although the forests are dense and of coniferous species, it is only in excessively dry seasons that flagrant fires are common.

As outlined in these factors, the regions of this country most seriously affected by recurrent forest fires are in the coniferous stands of the Northeast, principally in Maine, New Hampshire and the Adirondacks; the Lake States, and the dense stands of the Northwest covered by Douglas fir, western red cedar, western white pine, hemlock and lodgepole pine.

Grass fires are very common in the southern pine forests, in the western yellow pine stands of the Southwest and in the sugar pine-yellow pine stands of California. Although not resulting directly in the death of whole forests, they do considerable and irreparable damage by destroying all reproduction and young growth, and by eating away the base of trees until they are blown over by the wind or are attacked and ruined by insects or fungi. Even our hardwood stands are seriously injured by the disastrous work of forest fires. Although pure hardwood forests are never affected by crown fires, the common spring and autumn fires in this class of timber have done an immense amount of damage. This is especially true in the case of our woodlots and small holdings of the Ohio Valley and Middle West, as well as in the more extensive hardwood forests of the East, southern Appalachians and the lower Mississippi Valley.

Causes. — Ever since the first settlement of this country, our forests have been susceptible to severe fires due to a great variety of causes. Even before colonial days fires did considerable damage in our American forests. The early settlers often caused many fires in the clearing of land and no attempts were made to check them. The Indians have started many fires, especially in the West, to drive out game to facilitate hunting. Many grass fires, resulting later in serious forest fires, have been started in the South and West even up to present times with the hope of improving grazing conditions. Altogether about two-thirds of our whole forested area has been burned over at one time or another and much of this has been burned repeatedly.

Although it is very difficult to detect in every case the cause of forest fires, records have been kept which show that the most important causes of fires are railroad locomotives,

FIG. 25. — A CACHE OF FIRE-FIGHTING TOOLS READY FOR USE.

The shovel, axe and cross-cut saw are generally the best tools for fighting fires. Mattocks, rakes and forks are also used in making a fire line.

carelesss campers, clearing land and brush burning, lightning, hunters, and those of incendiary origin. It is a notable fact that of all the many causes of fires, lightning is the only one that is absolutely unpreventable. Burning to improve grazing, sparks from logging machinery, burning over blueberry patches and cranberry marshes are also important sources of fires.

Altogether, sparks from railroad locomotives and logging machinery, such as steam loaders, donkey engines and cableway skidders, have been the most serious causes of forest fires. Engines, especially on steep grades, send out great masses of sparks which readily ignite the grass or leaves along the right of way during the dry season. Railroad companies have suffered heavy damage suits as a result of this and many are now coöperating with state foresters and other officials in cleaning up their rights of way of all inflammable material, in installing effective spark arresters to prevent the emission of live sparks and very often in patrolling along their tracks during the dry summer season. Some have even adopted oil for use as fuel in mountain and forested regions. Several of our railroads have found the use of oil very economical especially in cheaper transportation of this fuel supply as against the heavier and more bulky coal. Some railroads, like the Chicago, Milwaukee and St. Paul, are electrifying their mountain divisions both to reduce the losses from forest fires and as an economical measure of operation. Several states now require by law the use of spark arresters, fire patrol and the cleaning up of the rights of way through forest regions. A few railroads have adopted certain remedial measures and have more than saved the increased cost of the installation and change of equipment in decreased law suits for fire damage. In New Jersey several of the railroads maintain plowed fire lines paralleling the tracks on each side, at a distance of one hundred feet. These fire lines are kept clear of leaves and grass so that in case sparks ignite the strip inside the fire line, the fire will stop when it reaches the plowed dirt, unless there is a high wind blowing.

Spark arresters consist of a heavy iron mesh intended to prevent the emission of large, live sparks. They are placed in the smokestack of the locomotive, and on steep grades, where they are needed most, they interfere to some extent with the draft, which means of course that it is more difficult to maintain a proper head of steam. In some cases firemen have driven large holes through the spark arresters or have detached them altogether so that constant inspection has been made necessary in some states to maintain the arresters in condition. However, it may be said that in spite of constant changes and improvements, no spark arrester has as yet been devised or perfected to suit all conditions and circumstances.

Careless campers constitute a very serious class in dealing with the fire question. This is especially true in mountain and forested resorts which are frequented by campers and visitors during the dry summer season, such as in Wisconsin, the lake district of Minnesota, the Adirondacks, and portions of the Rocky Mountains and the Pacific coast. Many a fire has been started by a lighted match carelessly thrown in the dry leaves. Burning cigars or cigarettes are also a considerable source of forest fires. Perhaps the most serious phase, however, is the neglected campfire. Too often a campfire is built against a large or old rotten log where it may smoulder for weeks. Other fires are built in the thick duff and are left in the morning or at night when a gust of wind may carry a spark a short distance and ignite the inflammable woods. The best way is to first build only a small fire and if it is impossible to find a site free from duff or litter dig a trench around it so the fire cannot run through the leaves or brush, and be sure to extinguish it on leaving it either for the day or for the night. Fires should always be built in the open and away from any standing or dead trees or any accumulation of dry, inflammable material.

Clearing farm land and burning brush, which always goes with it, is a common source of fires in our forest regions.

Many states, like Idaho and Massachusetts, now have closed seasons during which it is illegal to burn brush without a permit.

Lightning is a common source of fires, especially where thunder and lightning storms are prevalent. Fires started from this cause have been known to burn for weeks before breaking out into a serious fire. The popular belief that certain trees are more susceptible to being struck by lightning than others does not seem to be substantiated in actual fact. Lightning is especially dangerous when striking an old dead tree or stub. These ignite freely, especially during a dry time, whereas a healthy green tree may not ignite at all.

Our lodgepole forests in the Rocky Mountains have suffered greatly, in the past, from fires set by Indians to improve the hunting. In the South it is customary to burn over the woods every spring in order to improve the grazing conditions for the cattle and hogs. Even in parts of the West this is done to improve sheep grazing.

The question of burning over the woods every year to prevent the accumulation of a large amount of inflammable material has been agitated. This is called " light burning " and has as its object the annual disposal of the débris on the forest floor in order to prevent a severe conflagration during an unusually dry season. But this has proven to be poor policy because much injury results to the standing timber and reproduction, and in addition it is very expensive. This has been positively demonstrated in portions of California.

Effects. — The effects of forest fires, both direct and indirect, may be summed up in the following:

1. *Destruction of standing timber.* The greatest damage resulting from fires is the direct loss of millions of feet of valuable timber. During large conflagrations millions of dollars worth of standing timber is destroyed. If it were possible

to cut and utilize this burned timber, much of it could be salvaged, but usually other timber is being cut or facilities are not available to save the remaining standing timber. Contrary to the popular impression even severe crown fires do not actually consume the large trees. All the branches are usually burned off and the stem is severely charred, but if cut within a year or two after the fire, much timber can be

FIG. 26. — RESULTS OF A CROWN FIRE.

No living trees are left to re-seed the area and the wind soon blows over many of the trees left standing.

saved. Altogether, the greater part of our annual $50,000,000 fire loss can be attributed directly to the loss of standing timber. Surface fires also kill a great many trees, especially in our hardwood regions and southern pineries. Crown fires have done the greatest damage in the northern spruce and pine forests, the lodgepole pine stands of the Rocky Mountains and in the dense Douglas fir and white pine and hemlock forests of the Northwest.

2. *Indirect losses through windfall, insects and fungi.* Besides the enormous losses from the direct destruction of valuable live timber, forest fires do considerable damage by eating away the lower parts of tree trunks until the wind blows the tree over or permits the entrance of wood-destroying fungi. This is especially the case in the South where the longleaf pine is tapped for turpentine. Fires get into the exposed resinous faces and weaken the tree so that the first strong wind blows it over. Considerable losses have also been occasioned through the entrance of fungi and insects into fire-scarred surfaces of trees. Chestnut, lodgepole pine, western yellow pine and our oaks especially have suffered in this respect. Some trees, like Douglas fir, pitch pine and hemlock, either have thick barks or contain fire-resistant materials, such as tannin, so that they are protected by nature from injury by surface fires.

The longleaf pine seedling protects itself admirably from fires by its thick bushy crown. Even when this is burned off, the vegetative sprout at the tip will start up the following spring.

3. *Destruction of seedlings and young growth.* A considerable portion of our fire damage is the destruction of the young trees and reproduction which must be depended upon for the forests of the future. A dense young growth of conifers is especially inflammable in a dry season. Although it has no actual market value, our courts are beginning to recognize the potential or future value of young standing timber and are allowing damages in case of fires resulting from carelessness or negligence. In fact, under our present conditions of extreme fire hazard, the planting of new forests or expensive provision for natural reproduction is not warranted unless adequate fire protection is assured.

4. *Injury to the soil and streamflow.* As explained in the chapter on silvics, trees depend for their food and therefore

their growth and vigor on certain constituents in the soil, supplied from decayed leaves, twigs, etc., as well as from the mineral salts. If the humus and soil cover are destroyed, the soil is impoverished and a poor forest results. A severe fire will frequently burn out a thin soil down to the bare rock or at least to the hard mineral soil. This is especially the case in mountainous regions. It may take hundreds of years to bring back the soil to its normal condition and fertility. As the forest builds up the soil, it depends upon it for its best growth. In this same connection, it is apparent that after a fire the forest largely loses its function to prevent rapid run-off with consequent erosion of the soil, floods, etc. Many of our floods, especially in western Pennsylvania and the South, may be attributed to the work of fires as well as denudation from lumbering.

5. *Loss of life and general business.* It is estimated that several thousand lives have been lost in forest fires in the past fifty years. Whole towns have in some cases been wiped out of existence, together with many of their inhabitants. This has prevented in some cases the settlement of new regions by the loss of so much property. Even in recent years about seventy lives have been lost in forest fires on an average every year.

Many mountainous sections which have served as attractive summer resorts and recreation grounds have lost their attractiveness on account of fires. Even considerable game and fish have been destroyed by them. Many areas have reverted to the state through non-payment of taxes as a result of forest fires and this direct economic loss has been enormous in portions of the Lake States, especially Michigan, and in the South. For every dollar's worth of timber burned, it is estimated that several dollars in wages that would otherwise be expended in logging and manufacturing are lost.

Methods of Prevention. — In Europe the annual fire losses are less than one per cent of all the forested area, and all the forests are under some systematic plan of fire protection. In this country about one-half of our total forested area is

FIG. 27. — LOCATING AN INCIPIENT FOREST FIRE.

Lookout stations on peaks and high elevations for the detection of fires have proven to be effective means of controlling them. Telephone lines connect these stations with the nearest warden or ranger.

now receiving at least some attention in the matter of fire protection, whereas only a few years ago less than two per cent received any protection.

Our National and State Forests are now being organized under efficient plans of fire protection. Even the lumbermen and timber owners are uniting in protective associations and providing for a definite system of fire prevention. Large tracts of timber are now being successfully protected at an annual expenditure of from two to four cents per acre. Some associations of private owners have spent as high as ten and fifteen cents per acre in protecting their holdings against fires and they consider this as excellent forest fire insurance.

The principal methods of fire prevention include the construction of permanent improvements, such as roads and trails, to facilitate the work of getting over the forests and in stopping the spread of fires, lookout points and observation towers, and telephones connecting them with the nearest town and fire warden. In dangerous seasons fire guards or patrolmen are engaged to patrol the areas and watch for fires from lookout points or towers. In the clear atmosphere of the West fires have been detected from fifty to seventy-five miles distant. In the East fires have been detected up to thirty miles away. Railroads are building fire lines and clearing up their rights of way, and spark arresters and efficient ash pans are being installed where oil is not used for fuel. The occurrence of inflammable brush and slash after logging operations has been a serious menace in protecting forests from fire. The United States Forest Service now requires the disposal of brush on most of the timber sale operations on the National Forests. Piling and burning the brush has given excellent results. In some forests lopping and scattering the brush is required. When this is done, the brush soon rots and the damage from fires is greatly decreased. New York and Minnesota now require by law the disposal of brush by lopping or piling even on private logging operations.

By eliminating the inflammable grass, grazing has also proven to be an efficient aid in the prevention of fires, especially in portions of the West and South where grass fires are common.

Methods of Control. — The method of controlling forest fires varies almost with every large fire. In general, however, small surface fires can be beaten out with wet brush or a wet blanket or by simply throwing dirt on them. Large surface fires are best controlled by cutting out a fire line and trenching to the mineral earth well in advance of the fire. They are best fought just over the brow of a hill and at night when the atmosphere is damp. Some of these lines must be many miles in length to effectively check the progress of the fire. Crown fires are only fought by back-firing or waiting until they become surface fires. Under the presence of a high wind it is almost impossible to stop crown fires. The shovel, axe, saw and sometimes the rake and the mattock are the best tools used in fire fighting. Water can seldom be used owing to the difficulty of transporting it to the fire. Caches of tools, food and camp supplies distributed at strategic points over the forest are now being established and used with great success on our National Forests.

Forest Insects.

Although working in a more insidious and less visible way, insects have done an enormous amount of damage to our American forests. It is estimated that the western pine beetle killed over one billion board feet of timber — practically all the mature trees in the Black Hills of South Dakota — in the ten years from 1898 to 1908, causing a loss of $2,500,000.

Our forests have been attacked rather sporadically by

these insects of which there are thousands of species. The most serious damage, however, has been done by a comparatively few species. Some only attack living trees; others confine their work to dead and dying trees, while still others do great damage to sawed lumber and various wood products. Many of the insects limit themselves to certain tree species. Very often dead or fire-killed timber would remain sound and could be used several years after dying, if it were not for the destructive work of beetles or insects of some kind. Sawed logs or pulp wood should not remain long in the woods in the summer or they may be riddled and ruined by wood-boring insects.

Most of these attacks have been fought and curtailed by felling the insect-infested timber; by using the logs for lumber or other purposes; and by burning the tops, bark and other parts that may be infested with the larvæ, eggs or even the adult. Infected timber should usually be felled and treated during the winter or at the season of the year when the adult, larvæ or eggs will be disposed of. Spraying with a poisonous solution, such as arsenate of lead, has been used with success in the East in controlling leaf-eating insects, but it is not applicable to large forest tracts.

Some of the most dangerous species of insects and their work are described as follows:

1. The gipsy moth (Porthetria dispar) and the brown-tail moth (Euproctis chrysorrhœa) were both accidentally introduced from Europe and have spread with alarming rapidity over eastern New England. They eat the leaves of both fruit and forest trees, with the exception of most of the conifers, and have done an immense amount of damage. Already about $7,000,000 have been spent in fighting them. Massachusetts has appropriated annually for a number of years over $200,000 to suppress them, but they are still spreading.

They have been held in check to some extent by spraying, by destroying the egg masses and by the introduction of parasites to feed on them. Strong efforts are now being made by the government authorities and the states involved to prevent the spread of these two insects to New York and the other adjoining states.

2. The southern pine beetle (Dendroctonus frontalis) is one of the most dangerous insect pests in the country. It first became noticeable in 1891 when it spread over West Virginia, Virginia and North Carolina, killing millions of feet of valuable spruce and southern pine. Since 1902 it has been found in various parts of the South as far west as Texas and has broken out in intermittent attacks in the extensive pine forests of the Southeast.

3. The spruce beetle (Dendroctonus piceaperda) has been prevalent in New England and New York since 1818 and has broken out in some places, destroying many square miles of valuable spruce timber. It has, however, been rather sporadic in its attacks. By cutting out infested bodies of timber, it is estimated that on one tract of timber there was an actual saving of over $100,000. Another species of this beetle has done great damage in the Engelmann spruce of the Rocky Mountains.

Other prominent examples of insect attacks are the locust borer which riddles both the heart and the sapwood of the black locust, resulting in discrediting this valuable tree for planting by railroads and others in some regions; the white pine weevil; the catalpa sphinx; the larch sawfly; the hickory bark beetle; the powder post beetle; and a great variety of others.

Fungi.

A number of fungi have also done considerable damage to both living and dead timber and in some regions enormous losses have occurred. Among the most serious attacks may be mentioned the following:

1. The chestnut bark disease (Endothea parasitica) was first discovered in 1904 near New York City and it has now spread over nearly the entire range of this valuable tree. It has killed every chestnut tree to which it confines its attacks, in some places, especially in southeastern New York, New Jersey, southern New England, and in eastern Pennsylvania. The fungus works in through the bark, encircles the tree just beneath the cambium layer and eventually kills the tree by cutting off the circulation above that point. Its seeds or spores are scattered by the wind, birds and animals and it is estimated that it has already done over $25,000,000 worth of damage. No effective means of checking its spread has as yet been discovered and it is problematical as to whether or not it will eventually kill all our native chestnut trees. In forest management, where the disease is prevalent, other species should be encouraged as against chestnut.

2. Trametes pini attacks practically all the important conifers and it is a very common and serious disease. It enters the wood through wounds or fissures in the bark and causes a destructive rot in both the heartwood and sap. It is spread in the same manner as described above. No means have been devised to fight it but when discovered, diseased trees should be cut and removed if possible. This will result in greatly lessening the liability of loss from this cause.

A number of others of the Polyporus family, the white pine blister rust and others, have done great damage to our forests.

Grazing.

Grazing may be classed as an injurious agency when not properly regulated. Goats and sheep as well as cattle may prevent the growth of reproduction either by tramping them under foot or by nipping off the tender young shoots. Our woodlots especially have suffered in this respect. Whenever cattle are turned into a wooded area, the young growth is soon destroyed. Sheep and goats have also done a large amount of damage to our forests in the West. On our National Forests, however, this matter is being properly regulated and in some cases grazing may be used to advantage in stirring up the soil and enhancing the chances for a successful germination of seeds in reproducing forests. Woodlots should be devoted either wholly to pasturage purposes or to the growth of wood products.

Sand Dunes.

Blowing and shifting sands have done considerable damage on our eastern and western coasts, as well as in other portions, by covering up young or even mature forests, as well as valuable farm land and other property. They can be fixed and held in place by the planting of grasses and trees. France has expended many millions of dollars in the fixation of sand dunes in Gascony.

Miscellaneous.

A number of other injurious agencies have called forth certain protective measures in forest management. Among them may be mentioned windfall, frost and sun cracks, damage from snow and ice storms, injurious effects of smoke or gases and the attacks of animals. For instance, porcupines have killed thousands of lodgepole pine trees by girdling them at the base. Deer often nip off the ends of tender young seedlings, and rodents commonly preclude the possibility of reproduction by consuming all the tree seeds in certain localities.

Most of the causes of injury mentioned may be properly treated in the silvicultural management of forests. As an example, where trees are susceptible to windfall, the stand should not be sufficiently opened up or trees left standing alone so that they will be blown down by the wind.

Early or late frost may do much harm to certain species, especially in the seedling or young stages, both in freezing them and in heaving them out of the soil. The following species are especially susceptible to frost killing: black walnut, hickory, ash, beech, catalpa, a few oaks, fir and western yellow pine. Among the frost-hardy species may be mentioned birch, poplar, willow, maple, spruce, redwood, red cedar and Scotch pine.

Frost cracks are long splits caused in the stem of trees by the contraction of wood due to severe winter cold. They are followed by long projecting ribs on the tree, called frost ribs. A sudden fall of temperature is especially harmful. Hardwoods with large medullary rays are most susceptible, such as oak, beech and sycamore, as well as elm, ash and chestnut.

BIBLIOGRAPHY.

Brown, Nelson C. Methods of Forest Protection. Hardwood Record. Feb. 25, 1914.

Graves, H. S. Protection of Forests from Fire. U. S. Forest Service Bulletin 82.

Greeley, W. B. Better Methods of Fire Control. Proc. Soc. Am. Foresters. Vol. VI, No. 2, 1911.

Hitchcock, A. S. Methods Used for Controlling and Reclaiming Sand Dunes. U. S. Bureau of Plant Industry, Bulletin 57.

Hopkins, A. D. Insect Injuries to Forest Products. U. S. Bureau of Entomology Circular 134.

Long, W. H. Effects of Forest Fires on Standing Hardwood Timber. U. S. Forest Service Circular 216.

Leavitt, Clyde. Forest Fires. Report of National Conservation Commission. Vol. II, Senate Document 676, 60th Congress, 2nd Session.

Marlatt, C. L. Danger of the General Spread of the Gipsy and Brown-tail Moths through Imported Stock. U. S. Dept. of Agr. Farmers' Bulletin 453.

Metcalf, Haven, and J. F. Collins. The Control of the Chestnut Bark Disease. U. S. Dept. of Agr. Farmers' Bulletin 467.

Plummer, F. G. Forest Fires. Cause, Extent and Effects. U. S. Forest Service Bulletin 117.

Schlich, Wm. Manual of Forestry. Forest Protection.

Snyder, T. E. Damage to Telegraph and Telephone Poles by Wood-boring Insects. U. S. Bureau of Entomology, Circular 134.

Von Schrenck, Herman, and Perley Spaulding. Diseases of Deciduous Forest Trees. U. S. Bureau of Plant Industry. Bulletin 149.

CHAPTER VIII.

FOREST MENSURATION.

General. — Forest mensuration may be defined as the measurement of logs, trees and standing timber and the determination of growth and yields. It is not concerned with the measurement of any form of lumber.

FIG. 28. — MENSURATION CREW AT WORK IN SOUTHERN YELLOW PINE FOREST, TEXAS.

All of the trees are measured on strips of a given width and per cent, and these results are taken as an average of the whole forest.

Under the present economic conditions the measurement of the forest crop is of increasing importance, since with the rise of stumpage values, accurate estimates of the contents of timber tracts are necessary. In addition many of our progressive lumbermen and pulp manufacturers are now figuring on sustained yield from their forest land, so that the study of growth and yields is also of importance. Until the last few years it was customary to buy large tracts of timber on the ocular estimate of an experienced cruiser. Now, progressive lumbermen insist on a fairly accurate survey, as they can not afford to risk the chance of large error. Consequently mathematical methods are replacing the old ocular estimates, and while at present volume tables and yield tables are by no means complete much progress has been made.

Units of Measurement and Equivalents.

There are many units used at the present time in measuring timber. Among them may be mentioned the board foot, the cubic foot and the cord foot. Of these the board foot is the unit most commonly used.

A board foot consists of a board one foot square and one inch thick. This unit has attained its wide use in America since it gives an idea of the amount of lumber actually used, but it is by no means scientific or accurate when the entire contents of a log are used.

A cord foot is a stack of wood 8 feet long, 4 feet high and 4 feet wide. It contains 128 cubic feet gross, but, on account of the air spaces between the component pieces, may range anywhere from 70 to 102 solid cubic feet of wood. Various factors affect the solid contents of a cord. Split wood takes up more room than the same amount before splitting. Long limbs permit of more air space on account of the crooks. Logs of large diameter permit less air space between them, and finally since wood is apt to shrink in seasoning up to 12 or 14 per cent, the bulk of a long rank of cordwood may be diminished by that amount after standing one season in the woods.

Pulp wood, excelsior, shingles and heading bolts are the kind of materials which are mostly sold by the cord, and the dimensions may vary somewhat for these different products. For instance, in some parts of the country a cord may be 5′ × 4′ × 8′, making altogether 160 cubic feet, gross.

In addition we have various forms of short cords and stove wood. Either 18-inch to 2-foot lengths may be used and the price varies according to the solid contents in the cord. A standard cord is considered to yield from 350 to 600 board feet, depending on the size of the logs, with 550 board feet as the average.

The cubic foot is not widely used at the present time in the United States, being confined largely to measuring mahogany and other precious woods. With the increase of stumpage values, however, it is generally believed that the cubic foot will be a more common unit since the timber owner will insist on being paid for all this stumpage rather than make a generous allowance for slab and poor manufacture as has been done in the past.

The following equivalents are generally recognized by the Forest Service:

1 cord of wood 4′ × 4′ × 8′ = 500 board feet
1 railroad tie 7″ × 9″ × 9′ = 33⅓ board feet
1 cubic foot = 6 board feet
1 post, 4″ × 5″ × 7′ = 3 board feet

For converting foreign units the factors are as follows:

1 cubic meter = 35.316 cu. ft. or 1.308 cu. yds.
1000 cu. ft. sawn lumber = 2.36 cu. meters sawn lumber
1 cu. meter per hectare (2½ acres) = 14 cu. ft. per acre

Log Rules.

Log rules are tables giving the contents in board feet of logs of different diameters and lengths. Between 40 and 45 different log rules are used in the United States and Canada, since different rules are favored by the lumbermen in various sections.

The Doyle rule, for instance, is the official log rule in Louisiana, Florida and Arkansas, while the Scribner rule has received state sanction in Minnesota, Idaho, Wisconsin and West Virginia. The Spaulding rule is used in California. The Drew rule is used in Washington, and Vermont and New Hampshire each have a state rule.

The log rule was first mentioned in the lumbering literature of the United States in 1825 and the formula given at that time was practically the same as that of the Doyle rule which came out between 1870 and 1880.

Methods of Construction. — Log rules are generally computed in one of two ways, either by formula or by diagram, although good rules based on mill tallies have been made. Of the formula rules, the Doyle, Champlain and International are the most prominent and among the important diagram rules, the Scribner and the Maine rules may be mentioned.

The Doyle rule is made on the following formula: Deduct 4 inches from the diameter of the log at the small end for waste, square one-quarter of the remainder and multiply by the length of the log in feet. This will give the contents in board feet. The allowance of 4 inches is for slabs, sawdust, edgings, etc., and it is extremely liberal for small logs, making the Doyle rule highly conservative for any but large sized logs.

The International rule is one of the best from the standpoint of theoretical construction, but so far it has not been widely accepted by the lumbermen. It is designed to give volume of logs cut with a band saw, cutting one-eighth inch kerf, and to allow for shrinking, seasoning, waste and normal crook. The rule was scientifically computed by mathematical formula and has been checked by mill measurements.

The Scribner is by all odds the most important of the log rules based on diagrams. It was made by drawing a diagram to scale, showing the width of boards that could be cut from logs of different sizes, the diameters being taken at the small end inside the bark. It gives fair results for small logs, but for large logs over 28 inches in diameter the results are too small. However, it was figured that defects on over-mature logs might be allowed for in this way. The Scribner rule — as Scribner Decimal *C* (*i.e.*, calling 104 board feet 10 and 110 board feet 11, etc.) — has been adopted by the Forest Service for use in timber sales and has given satisfaction.

The Scribner rule has been combined with the Doyle rule to make the Doyle-Scribner. Doyle values are used up to 28 inches and Scribner values above, using the most conservative portions of both rules. As a result the amount actually obtained by sawing up the log may "over-run" the log scale considerably, to the profit of the manufacturer.

The Maine or Holland rule is also based on diagrams and allows for boards down to 6 inches in width. It is designed for small logs and is very satisfactory for them. However, it leads to wasteful lumbering if applied to long logs, as a jobber or contractor will cut in lengths favorable to him regardless of the waste that ensues. The Maine rule is considered the best practical rule in use.

A third classification might be made with standard rules, in which the logs to be measured are compared to the contents of a log of a given size. The New York Standard Rule is the best known of this type and is quite commonly used in the spruce operations in the Adirondacks. In this case the unit is a log 13 feet long and 19 inches in diameter at the small end. This standard or "market" contains about 192 board feet and the converting factor used in the Adirondacks is 5 markets or standards per thousand board feet or 3 markets to the cord. To reduce a log of any dimensions to standards, its length and diameter at the small end squared are compared to the dimensions of the standard as follows:

$$\text{Volume in standards} = \frac{D^2}{19^2} \times \frac{l}{13}.$$

This log rule is quite accurate in measuring pulp wood or wherever the entire volume of the stick is used.

The New Hampshire or Blodgett rule is based also on a standard, the unit in this case being a log 1 foot long and 16 inches in diameter. The statute sanctioning this log rule states that there shall be 1000 board feet to every 100 Blodgett feet. This, however, does not hold as the lumbermen find that there are 115 New Hampshire feet per 1000 board feet when the diameter is taken from the middle of the log, and 106 when taken at the small end. The formula used when converting logs into New Hampshire feet is as follows:

$$V = \frac{D^2}{16^2} \times \text{length}.$$

In converting a standard foot into board feet it is not possible to get an accurate result by using the same equivalent for all logs regardless of size. Small logs yield considerably less lumber when sawed than large logs owing to larger proportion of slab and sawdust, hence different factors should be used.

Relative Values and Discussion. — Widely varying results may be obtained by scaling the same skidway with different log rules, owing to the different values that are given by different scales for logs of the same size. A brief comparison of the various log rules in use can be made as follows:

CONTENTS IN BOARD FEET OF 16-FT. LOGS OF DIFFERENT DIAMETERS AS GIVEN BY VARIOUS RULES.

Rule	Diameters			
	6″	16″	26″	36″
Scribner	18	159	500	923
Doyle	4	144	484	1024
Holland (or Maine)	20	179	507	1026
Cumberland River	...	121	320	685
Derby	28	195	512	977
International	20	200	555	1085

Log Scaling. — Log scaling is the measuring of logs by a given log rule or log scale. It is generally done by one man although two men may work together. The scale stick is applied to the small end of the log inside the bark, the length of the log is ascertained and the contents, as given by the rule used, are recorded. The scale rule is generally a straight piece of hickory tipped with a plain binding of brass or iron and is marked with figures, showing the contents of logs of different diameters and lengths. Sometimes a caliper scale is used where the diameter is measured at the middle of the log. Under these circumstances the jaws of the calipers are generally sprung slightly to make a deduction for the bark. This deduction, of course, is the same for trees of all ages and thicknesses.

The lengths are taken ordinarily in standard even lengths of 8 to 20 feet and an overrun of from 2 to 5 inches is made in order to allow for the loss by checking, as the log seasons, or "brooming" in the drive. In some cases there is an enormous waste of valuable lumber owing to excessive allowance. For getting the length of the log, a pole or tape is commonly used, although a wheel having 10 spokes, 6 inches apart, is

FIG. 29. — FOREST OFFICER SCALING DEAD LODGEPOLE PINE — ARAPAHO NATIONAL FOREST, COLORADO.

In each of the 6000 timber sales made yearly, the Forest Service requires the measurement of every piece of lumber sold.

sometimes used. Every time the colored spoke of the wheel, running along the log comes down, 5 feet has been measured and the length of the log may be easily ascertained within 6 inches. An experienced scaler can estimate by eye the lengths quite accurately. In scaling, the diameters are ordinarily rounded to even inches and a log 9.8 inches in diameter would be scaled as a log of 10 inches, while a log 11.4 inches would be scaled as an 11-inch log.

The results to be obtained in scaling are very largely dependent upon the personal equation and the care exercised by the scaler. Some scalers are extremely conscientious and will endeavor to measure the length of each log, while others may guess the contents of a log without using the scale stick at all. Two conscientious scalers working independently may get different results on the same skidway when the same log scale has been used, the difference being due to allowance for defects, etc. The cost of scaling ordinarily ranges from five to fifteen cents per thousand board feet.

Volume Tables.

A table of contents for any one species, showing the number of board or cubic feet in trees of different sizes, is called a volume table. The contents given are average only and while they will hold true in computing the contents of an entire stand, they may not give a true result for any single tree.

Kinds. — Volume tables are of different kinds, the most important of which are:

First. Volume tables based on diameters alone.

In this type the contents are considered to be a function of diameter only. It is the simplest form of volume table, easy to make, and ordinarily the first to be used. When not used outside of the region for which it was intended such a table gives fair results.

Second. Volume tables based on diameter and height of the tree.

Under this heading may be grouped those volume tables in which the contents for trees of different total heights or for different used lengths are given. There are several forms of this type of volume table, such as those based on diameter and merchantable lengths and the volume table based on diameter and tree classes. The most widely used, however, is the volume table based on diameter and classified heights.

Third. Graded volume tables.

This table is the only one which shows quality as well as quantity produced. As the name indicates, such a table gives the amount of lumber of each grade that can be sawed from trees of different sizes. A few such tables have been made and when care is taken in constructing them, they are extremely valuable since they give the exact value of each tree as it stands in the woods. On account of the fluctuation in price and grades and the variation in sawing practice, such tables need constant revision and can be used effectively only with valuable woods and where intensive management prevails.

Construction. — The graded volume table is made by following a tree which has been measured and numbered through the mill, log by log, and recording the amount of timber of each grade that is obtained. The contents of all trees having the same diameter are averaged and any irregularities are evened off by a curve.

The volume table, based on diameters alone, is made by scaling a large number of trees (500 will serve if taken with care but 1000 trees are better) and then plotting the contents on cross-section paper, diameter on age. A curve is then drawn to strike an average and the values are read from the curve.

The table based on classified heights is rather more complicated but the underlying idea is the same. The data is best secured on a large lumbering operation where the large number of trees felled permits the rejection of any specimens that deviate from the normal. A crew of three men is most effective and the equipment needed consists of calipers, tape, scale stick and notebook or tally sheet.

The steps are as follows:

First. Scale 500 to 1000 trees, using the log scale in common use in the region. Volume tables based on different log scales will show different results, so it is generally better to use the log rule favored by the local lumberman unless there is good reason for changing.

Second. Group the trees by diameters and classified heights, *i.e.*, keep records of all of the 6 inch, 40 foot and all 7 inch, 40 foot trees together, etc.

Third. Plot the contents of the trees. — Volume on diameter in height classes, *i.e.*, make a separate curve for the 40 foot trees as well as for the 45, 50 and 55 foot trees, etc.

Fourth. The values that are read from this curve may increase irregularly so they are plotted again; volume on heights in diameter classes. The final values may be read from these curves and will ordinarily increase from size to size in a uniform manner.

Use. — In computing the contents of a forest, following a valuation survey, volume tables are practically indispensable. The tally sheet of the estimator will give the number of trees of each diameter and species and it is then a very simple matter to multiply the number of 10-inch spruce trees, for instance, by the value given in the volume table for a tree of that size. The computations are comparatively short and the results much more accurate than those obtained by using a rule of thumb or from ocular estimates. If the table used is based on classified heights, a curve, height on diameter, must be drawn to show the average heights for trees of different diameters, in order that the proper values in the tables may be used.

Practically all systematic timber cruising is based on volume tables, and by their use a well-trained forester can go into a new region and compete successfully with local woodsmen whose methods of estimating are based almost entirely on local experience.

Cruising.

Cruising is the woodsman's term for estimating standing timber. In most cases it is rarely an actual measurement of the entire timbered area, but a portion is generally measured and this fraction is taken as typical of the whole. The estimating systems may be classified as follows:

First. Ocular.
Second. Systematic.

First. Ocular. The practice of estimating timber by eye is extremely primitive and is the first method that is used in any virgin country. It is an extremely rough method and its accuracy depends entirely upon the care and experience of the estimator. In the past, ocular estimates served very well because generally a highly conservative estimate was made and due allowance for defects provided for, but with the increase of stumpage values a more accurate valuation must be obtained and, consequently, the day of ocular estimates, on a large scale, is passing.

Ocular estimates were either absolute guesses or else a rough count was made of the trees and sample trees selected. With an experienced cruiser good results may be obtained in this way, but the same cruiser outside of his own type of forest would be apt to get inaccurate results, as his experience would be of no avail.

Second. Systematic Methods. Under present economic conditions and with the need for accurate estimates of growth and yields, more than a mere guess is needed. For this reason actual measurement of a definite portion of the forest is fast replacing the rough estimate based largely on local experience.

Systematic cruising is done by measuring trees on a strip or by selecting sample plots which are typical of the forest as a whole. In the strip method a compass line is run across the tract to be estimated and all of the trees on a strip of a certain width are measured. When the opposite side of the tract is reached, an offset is made, the length of which depends on the per cent to be covered, and then a line is run back across the tract. By crossing the forest repeatedly, from 2½ to 25 per cent of the forest may be measured and an accurate idea of the condition of the topography of the tract, timber and lumbering possibilities may be obtained.

The instruments used in cruising are compass, chain, calipers to measure the diameters and a hypsometer to ascertain the heights. Under certain circumstances distances may be measured by pacing and heights estimated by eye. A four or five man party is the ordinary crew, and the duties are as follows: The compass man runs the course, takes notes on the topography, drainage, type of forest, and records the distance to roads, streams, etc., and keeps a rough sketch map during the survey. The leader of the party generally acts as tally man and records the measurement of the trees, (a separate record is kept of the trees as obtained by the caliper men for each acre or type). Two men caliper the trees 4½ feet from the ground and if a fifth man is available, he will assist the tally man or the compass man and meanwhile gathers data on heights, types, condition, etc.

The strips should always be run up and down the slopes in order to get a true average of the stand. All roads, streams, ridges should be noted and all data recorded which will have a bearing on lumbering and silvicultural methods of management. The ordinary valuation survey consists in measuring the trees on a strip 4 rods wide, each caliper man covering a distance of 2 rods out from the compass man's line. In open timber, like western or southern yellow pine, a strip 10 rods wide may be covered, but under these circumstances all trees on the strip are counted rather than measured and every fifth tree is tallied for contents.

The distance covered by an estimating crew depends largely upon the topography, type and methods of getting diameters. In a dense spruce forest of Maine and Northern New York, a crew can make $2\frac{1}{2}$ miles a day with diligence. In ordinary hilly country with timber of ordinary density from 3 to 4 miles a day is considered a good day's work, while in very open country with sparsely timbered areas five or six miles of line may be run in a day.

In locating sample plots average acres may be laid off according to one of two methods. The plots are located either arbitrarily in the different types composing the forest or they may be located on a line across the tract, at a fixed distance apart. The plots may be either round or square, accurately measured or approximated. All of the trees may be calipered or the number of trees on a given plot may be counted and the contents obtained by a rule of thumb. The best rule of thumb in practical use is the one used by Ward in connection with his sample plot method. His method consists in counting all of the trees on a plot with 118 feet radius (one acre), and selecting an average tree by eye. Its merchantable length is estimated in terms of 16-foot logs and the diameter inside the bark at the top and bottom of the stick to be used is estimated. The average diameter inside the bark is then obtained and the contents of the trees computed as follows:

(Mean. diam.^{2} − 60) .8 = Contents of average 16-foot log in board ft.

Multiplying the number of logs in the tree by this figure will give the contents of that tree. With this figure as an average, the contents of the sample acre may be obtained by multiplying by the number of trees standing upon it.

Determination of Contents of Trees.

For ascertaining the contents of trees in board or cubic feet, either for commercial or scientific purposes, a quick method is desirable. In logging work this is very easily obtained by scaling each log by means of a scale rule which immediately gives the contents in board feet.

In ascertaining the cubic contents of a tree it is necessary to use one of several methods. The customary way is to multiply the average of the basal areas of both ends of the log in square feet by the length of the log in feet.

In some cases it is desired to know not only the present contents of the tree but the contents ten years ago, twenty years ago, etc. Under these circumstances a stem analysis is made. A complete or tree analysis comprises the following measurements: Diameter breast high outside the bark; height of stump; length of each section; diameter in and outside of the bark at each cross-section; diameter growth at each section; age and width of sapwood; total age of tree; total, clear and merchantable length.

In making stem analyses only sound trees should be selected as any inferior trees may lead to erroneous conclusions. Under ordinary circumstances the easiest way is to follow the fallers on a logging job and measure the logs which they cut.

Cross-cuts should be made with a saw and these should be at right angles to the axis of the tree. The diameter growth of each section is studied by counting the annual rings, the character of investigation determining whether the count is started at the centre or the outside If the average diameter growth is the object, count the rings out beginning at the pith. When volume growth is wanted, count the rings in and measure out by decades; that is, by making a mark at each tenth ring and measuring the distance from decade to decade can readily be computed the diameter of the section ten years ago, twenty years, thirty years, and from this the volume of the log may be computed by decades.

The number of rings at each cross-section indicates the age of that section and by computing the volume of the tree, considering both height and diameter, the contents may be easily ascertained. Such an analysis is indispensable in an accurate growth study.

Where a record of growth is not necessary but a statement of present dimensions will suffice, taper tables may be used in place of stem analysis. Under these conditions the dimensions of the trees are recorded at certain fixed distances above the stump, so that by means of a log rule the contents of the total tree may be worked up, but without, of course, considering age or growth. Taper tables will be sufficient for constructing or amending volume tables but since age is not considered, *i.e.*, no rings are counted, they are not of use in the study of growth.

Growth Studies.

By growth is meant the increase of a tree in size. Height, diameter, area and volume growth cover the phases of a tree's increase. Of these volume growth is the most important, considering it from an economic standpoint.

The methods of studying diameter growth and volume growth were touched upon in connection with stem analysis, but the study of the growth of a single tree is of minor importance in comparison with the volume growth of stands.

FIG. 30. — STUDYING THE GROWTH OF BEECH IN THE CATSKILL MOUNTAINS.

A crew is at work making a stem analysis to determine the rate of growth. A large number of trees are felled and measured and the data averaged together.

As previously described, the volume growth or increment of a single tree may be ascertained by computing its contents and comparing them with the contents of the tree ten, twenty and thirty years ago, but this method involves a maximum amount of computation and is chiefly valuable in scientific investigation.

For practical purposes the increment of forests is generally calculated or predicted in one of two ways. The mean annual increment method is based on the assumption that a forest will grow at the same rate during the next five years that it has maintained up to the present time. For example, a pine forest at forty years contains 24 cords of wood per acre; the mean annual growth, therefore, is .6 cord per acre per year. According to this method the increase during the next ten years would be at the same rate and at fifty the stand should yield 30 cords per acre. The accuracy of this method depends entirely on whether the growth rate is increasing or decreasing.

The growth per cent method is also used in somewhat the same way and its use is largely confined to mature forests whose height growth has practically ceased and the forms of whose component trees remain unchanged. Schneider, a German forester, put forward a formula for determining the growth per cent, depending on the rapidity of the radial growth.

A sample plot is measured as the first step in this method and the average tree or trees selected. They are then chopped or bored breast high to ascertain the number of rings in the last radial inch. An instrument, called an increment borer, consisting of a hollow auger, is commonly used to extract a core whose rings can be counted. The growth per cent is then found according to the formula

$$\text{growth per cent} = \frac{400}{nD},$$

where D equals the diameter $4\frac{1}{2}$ feet from the ground and n equals the number of rings in the last radial inch. If the stand runs 8000 board feet per acre and the growth per cent is found to be 3, then the forest is increasing at the rate of 240 board feet per year and in five years the stand would run 9200 board feet per acre.

Both the above methods give results which may serve the purpose where only approximate figures are needed. For scientific purposes more accurate methods must be used.

Yields.

In predicting the contents of a forest at specified ages a yield table is used. This is a tabular statement of the contents per acre of a forest growing on a specified site and handled in a certain manner. The best tables contain yields for each of the three qualities of forest soil and, in addition, state whether the forest is pure or mixed, thinned or unthinned.

Yield tables to give satisfactory results should only be used in the region where they are constructed. Even in Germany with its limited area in comparison with the United States "local" yield tables are preferred to "general" tables.

Two kinds of tables are recognized.

1. Normal.
2. Empirical.

A normal yield table shows the yields per acre for fully stocked stands. It gives the average maximum yield actually obtainable, provided the stand has not been injured by fire, wind, insects, etc.

Empirical tables show the average stand per acre for the whole forest. They consist merely of a table of average contents and in contrast the normal tables are indefinite and inexact.

The selection of plots for making tables requires considerable skill and the general tendency is to pick stands which are understocked. Complete measurements of the stand are taken and the contents per acre are plotted on cross-section paper — volume on age.

Use of Yield Tables. — Yield tables are extremely valuable in determining the value of second growth timber, to predict future yields and as a check in estimating the contents of stands. With the increasing importance of exact information concerning forest management, yield tables are becoming more necessary and the next decade will no doubt see the number of American yield tables greatly increased.

Working Plans.

Any well-managed forest should be cut according to a definite plan and where an equal amount is to be cut each year, the cut being exactly equal to the growth (sustained yield), a detailed plan covering the needs and treatment of each portion of the forest is necessary.

The German forester is very explicit in describing his forest and defining the exact manner in which it should be cut during the next cutting period or even rotation. In American forest practice, less intensive, on account of different economic conditions, the working plan document may discuss quite briefly the procedure for putting the forest in the most productive condition.

The ideal working plan considers present condition and growth with possible improvement of both, as well as protection against any attacks, etc., for each forest type. With a forest map showing types, topography and stand as a basis, a brief description will suffice, so that a well-trained woodsman may carry out the purpose of such a plan and by starting the cuttings in the portions of the forest that are damaged, or over-mature, and consequently slow-growing, the entire forest may be rejuvenated and put in splendid growing condition.

BIBLIOGRAPHY.

CARY, AUSTIN. Manual for Northern Woodsmen. Harvard Univ. Press.

CHAPMAN, HERMAN H. Timber Estimating. Proc. Soc. Amer. Foresters, Vol. IV, No. 1.

DANIELS, A. L. The Measurement of Saw Logs and Round Timber. Forestry Quarterly, Vol. III, No. 4, 1905.

GRAVES, HENRY S. Forest Mensuration. John Wiley & Sons, New York City.

GRAVES, HENRY S. Scaling Logs. Forestry Quarterly, Vol. III, No. 3, 1905.

TIEMANN, HARRY D. Methods of Discounting for Defects in Scaling Logs. Forestry Quarterly, Vol. III, 4–339.

CHAPTER IX.

LUMBERING.

History.

The history of the lumber industry in this country dates back to 1631 when the first sawmill was established at Berwick, Maine. This was a rough affair, driven by water power and comparable to the common portable sawmill of to-day which cuts about 3000 to 6000 board feet per day. Since this time, the lumber business has taken wonderful strides until, at the present time, it ranks as the third in importance among the industries. In this country the output of lumber exceeds that of any other nation. Some of the modern mills now turn out as much as 700,000 board feet of lumber in a day.

The native forests of this country were more abundant and of greater variety than those of any other country. At first only the finest white pine along the Atlantic Coast was cut for lumber purposes. Immense quantities of good timber were cleared off and burned up to make way for our farms, and until recent years it was generally thought that our timber resources were practically inexhaustible. For a long time Maine was the center of the lumber industry. Squared white pine and spruce logs and ship timbers were sent to all parts of the world. New York was next opened as a great source of our lumber supply and until 1850 was the leader in lumber production. Pennsylvania was the temporary center next, and around 1880 Michigan held the leadership. Up to this time white pine was the greatest source of our lumber supply. But the Lake States were rapidly cut over and the center of the industry next shifted to the vast undeveloped stands of yellow pine in the South. The lumber production increased in leaps and bounds to meet the de-

FIG. 31. — A STEAM LOG LOADER IN ACTION.

Labor-saving devices are being constantly introduced in the lumber industry to reduce the number of men and horses employed. This Barnhart loader runs on its own track and can load up to 200,000 board feet of logs in a day.

mands of the increasing population and the development of the middle and far West. At the present time more long-leaf pine is cut than any other species. In fact, it composes along with the other southern yellow pines about one-third of our total lumber cut. But even the extensive bodies of southern pine are becoming depleted and the industry is rapidly moving to the North Pacific Coast where the last virgin forests remain to be cut. At the present time Washington leads all other states in lumber production and the principal tree, Douglas fir, composes about twice the total amount of any other single species left standing in our American forests. In fact, about one-half of the total remaining supply of timber in this country is found in the five northwestern states of Washington, Oregon, California, Idaho and Montana.

Great improvements have also characterized the development of the lumber industry. In the early days only the finest and choicest specimens were selected for cutting in our forests. Many species, such as hemlock, red gum, sycamore, elm and other hardwoods, were left as worthless. The culling out of the white pine, oaks, etc., in the East left the forest to be reproduced from the inferior trees so that our present stands are relatively poor in quality as well as in quantity. Very often only one or two logs would be taken from the clear stem in logging and the remainder left in the woods to rot. Stumps were cut up to two or three feet in height.

In sawing methods, also, there have been great changes. Up to recent years the wasteful circular saw cut from one-fourth to three-eighths of an inch kerf, wasting from 10 to 20 per cent of the log sawdust alone. About the middle of the last century the gang saw and still later the band saw were introduced. These saws are great improvements in that they saw better lumber and do not waste so much in the form of sawdust.

Present Stand and Annual Cut.

It is estimated that the present stand of timber in this country, of all classes, covers about 550,000,000 acres and that this contains 2,500,000,000,000 board feet. The supply of virgin timber at the present rate of consumption will probably last, including growth, about thirty-five to fifty years, and from then on we will be forced to use second growth timber exclusively.

The annual cut of lumber in this country is about 40,000,000,000 board feet. About 78 per cent of the lumber cut is coniferous and the remaining 22 per cent is composed of hardwoods. Twenty per cent of the total remaining stand of timber left in the country is hardwoods. The principal trees cut in order of production are as follows:

Species.	Per cent.	Species.	Per cent.
Yellow pine	35	Hemlock	7
Douglas fir	14	Western pine	4
White pine	9	Spruce	3
Oak	8	Maple	3

Altogether about 150 separate species, to a greater or less extent, are used for lumber in this country. The principal lumber producing states at the present time are:

States.	Per cent.	Principal species.
Washington	11.0	Douglas fir, cedar, western yellow pine and Sitka spruce.
Louisiana	9.6	Southern yellow pine, cypress and cottonwood.
Mississippi	5.5	Southern yellow pine, red gum and oak.
Oregon	4.9	Douglas fir, western yellow pine and spruce.
North Carolina	4.9	Southern yellow pine, oak and yellow poplar.
Arkansas	4.8	Yellow pine, oak, red gum, and cottonwood.
Wisconsin	4.8	Hemlock, white pine, birch and basswood.

Practically every state contributes to the lumber cut, although the southern and western states produce the bulk of it now.

Rise in Stumpage Value.

The rapid rise in stumpage values of practically all of our timber trees has been no less than phenomenal. This in itself indicates in a forcible way the growing scarcity of our wood supplies and foreshadows the rapidly approaching time when we must place our forests under a definite system of

FIG. 32. — FLUME EMPTYING INTO LOG POND, WASCO CO., OREGON.
This flume is of box construction, 4 feet wide at top, and 2 feet deep.

permanent management in the same way that the European nations have done. Up to about 1860 timberlands did not even have any speculative value. Since that time, however, they have been on the rapid increase. In fact, holders of timber lands have made a great deal more money out of their investments than the lumber manufacturers, whose profits have not been great owing to increasing costs of labor and machinery, together with the intense competition that has existed in the production of lumber in the past twenty years or more. In brief, the majority of profits in the lumber business are due to the rise in stumpage values.

Perhaps the best example of the increase of stumpage values is found in the case of our best all around lumber tree, white pine. In 1860 to 1866, in Michigan, stumpage could be purchased for about $1 per acre. An acre often contained from 20,000 to 30,000 board feet. In 1880 white pine was worth about $3 per thousand board feet. In the same state and at the present time it is bringing from $10 to $20 per thousand board feet on the stump.

In the East twenty years ago hemlock did not command any price whatever, as it was not considered worth anything. To-day it is bringing from $3 to $5 per thousand on the stump and lumber which sold for $7 per thousand ten years ago is now bringing about $20.

Southern yellow pine stumpage has increased in the last ten years from $1 per thousand to about $4 per thousand.

The value of any stumpage depends upon its quality, accessibility and relative demands on the market. A great many timber tracts formerly considered of little or no value on account of their remoteness from a railroad or market are now becoming of considerable value because of improved market conditions and use of such modern methods of logging, as flumes, railroad logging, etc.

Methods of Logging.

General. — Logging methods are more varied and are developed on a greater scale in this country than in any other. American conditions, such as differences in topography, size of timber and inaccessibility, present so many difficulties that a great many different methods have been devised to get our logs to the sawmill in the cheapest possible way. With the gigantic redwoods and Douglas fir on the Pacific Coast, special logging machinery has had to be devised to haul in logs which were too heavy and unwieldy to be handled by horses or oxen. In the South cypress is usually deadened by girdling a year or more before cutting so the logs can be floated to the mill. In the swamps and in rough mountains,

FIG. 33. — SKIDDING LOGS BY STEAM IN NORTHERN HARDWOOD FOREST.

A Clyde ground skidder in action. Logs may be hauled in by a steel cable for distances up to one mile.

overhead cableways are stretched on which to bring in the logs to the railroad. Each logging operation presents difficulties of its own.

Up to the present time the lumber business is the only large industry which has not applied scientific methods in operation. The labor is largely unskilled and cheap and is made up of French-Canadians in the East, negroes in the South and miscellaneous foreigners in the Lake States and the West.

Camps are roughly constructed of logs, and are temporary in character except in places in the South where the men live in box cars which are transported from one camp to another on log trains. About 60 men make up the usual lumber camp.

FIG. 34.—SKIDDING LOGS BY STEAM DONKEY ENGINE IN NORTHERN IDAHO.

Chutes are frequently used to facilitate the movement of logs from the woods to the railroad landing. This timber was killed by a forest fire but it is still sound if logged in time.

Felling and Bucking. — These operations usually go together and the first consists of making an under cut with an axe on one side to direct the fall of the tree and then using a cross-cut saw to complete the cut. Formerly much valuable timber was lost in cutting high stumps especially in the West where spring boards were used to get above the flare of the stump. But lumbermen are now cutting low stumps and are saving this waste. After felling, the stem or bole of the tree is " bucked up " into logs of convenient length for hauling to the mill. Logs are usually cut in 16-foot lengths although they are cut up to 60 feet for long timbers in the South and West.

Skidding. — This is the operation of dragging the log to the landing where it can be transported to the mill by one of the many methods, such as driving, fluming, or the use of logging railroads, etc. It is seldom used for distances of over one-quarter of a mile. A pair of tongs or a chain is attached to the end of a log and it is dragged by a team to the landing or skidway.

Power skidding is now commonly used to drag the logs directly from the woods to a railroad track where they are loaded on cars and sent to the mill. A long steel cable up to a mile in length is used to drag the logs to the landing. This cable may be supported in the air from spar trees or laid out directly on the ground. A steam engine is used to haul in the cable with the logs attached.

Log Transport from the Woods to the Mill. — A great many methods have been used to transport the logs from the woods to the mill. Sometimes this distance may be 100 miles or more. In the old days driving was the most common method. Logs were rolled into a stream and floated to the mill. When the stream was too small, splash dams were built to assist them in their progress. This method is cheap but is useful only with conifers and a few hardwoods that will float. A great many logs are lost and they are delivered all at once in the spring when the freshets are running. This means that a sawmill can seldom run the year round.

Hauling on sleds in the winter is also a common method in the North and West. Loads of logs up to 15,000 board feet can be hauled at once. It is often used in connection with other methods, such as driving, flumes, railroads, etc.

Logging railroads to haul logs to the mill are coming into greater use every year. Although the initial expense for construction and equipment is very great, to be sure, it brings to the mill a steady and constant supply of logs; no logs are lost in driving and it can be used with heavy as well as light logs. It is especially used in the flat pine regions of the South and in the North and West where there are long hauls to the mill. There are 265 logging railroads in the State of Washington alone. When a large tract is cut over, many of these railroads are maintained to develop agriculture and mining as well as the timber resources of the region through which they run. Traction engines that run on snow by the use of an endless chain over the driving wheel, and haul in long loads of logs on sleds, are used in Maine and Minnesota.

Flumes are used in inaccessible regions where the use of railroads or driving is impossible or inadvisable. They are usually V-shaped and are used to transport logs and ties as well as lumber. They are chiefly used in the West. One flume in California is 55 miles long. They are expensive as a rule and depend upon a constant flow of water for successful operation. On long flumes the leakage is often so great that additional auxiliary streams must be turned in to float the logs.

Chutes and timber slides are used on steep slopes where horses or oxen cannot be used to advantage. They operate both by gravity and with the help of a team on level stretches. A slight application of grease is often used to facilitate the progress of the logs. They usually extend from a skidway to the stream or to a landing where they are loaded on cars or sleds. They are found principally in the Southern Appalachians and in the Northwest.

Rafting is employed on some of our larger rivers and on the Pacific Ocean. Enormous quantities of logs, usually in full tree lengths, are held together by chains and either towed or floated down the stream to their destination.

Methods of Manufacture.

General. — Methods of manufacturing lumber have been notable for the many changes and progress in the development of mill machinery.

FIG. 35. — FLOATING LOGS TO THE MILL.

Driving and rafting are cheap methods of transporting logs to the mill but they are gradually being displaced by the logging railroad which is more sure and can be used with heavy as well as with light species.

Demands for lumber have been so great in this country that very often mills built and intended to turn out 50,000 board feet per day have been later developed to turn out twice that amount. About 1890 the band saw was generally introduced with a great consequent saving in sawdust. All our large mills cutting from 100,000 to 700,000 board feet per day are now equipped with band saws. The gang saw which has been used for many years in Germany has also been introduced to turn out the product more rapidly, and is used in connection with band saws.

Not only is there more saving in sawdust but better lumber is being turned out with the modern machinery now in use. Our edgers, trimmers and planers are equipped to manufacture straighter and better lumber. Slabs and other mill refuse formerly wasted or sent to the burner are now used to make lath, box boards and in some cases a great variety of small wooden articles.

A few years ago only the best species were used for lumber purposes. Now, every tree of marketable size is utilized in this way. Examples of woods that have recently been introduced on our markets are: red gum, sycamore, beech, western larch, black or tupelo gum, cottonwood, etc. The trees are now being cut down to 6 inches in diameter at the top. White pine is sometimes taken down to 4 and 5 inches at the top.

Log Storage. — When the logs are brought in from the woods they are usually rolled from the cars or floated into a log pond where they are stored or kept until needed in the mill. These log ponds are usually only a few acres in extent and are sufficiently large to hold millions of board feet. In the past sawmills usually started cutting in the spring after the log drive, but now many sawmills run practically all the year and in the northern part of the country so-called "hot ponds" are maintained. They are kept from freezing by means of steam pipes around the pond. This permits the storage of logs in the winter time and the sawing of logs whenever they need them. In small and portable mills, logs are rolled down a slope to the log carriage and sent through the mill as they come from the woods. This is not possible in large mills where millions of feet are driven down a stream at once or long trains of logs are brought in from the woods.

Sawmill. — The modern sawmill is usually equipped with all the labor-saving devices possible and many are run by electricity so that each part of the machinery may be run and controlled separately. There is usually the large band saw, with a steam-driven carriage on which the logs are carried past the saw. The boards are then sent through the edger to clear off the uneven edges, and the trimmer to trim off the ends at the proper length.

Sometimes a gang saw will be used to saw up a squared log or "cant" into boards one inch in thickness. The boards and slabs are moved about on rolls driven by machinery to save manual labor.

The sawyer, who directs the log carriage against the saw and manipulates the log into position by means of a "steam nigger," is the highest paid and most important man in the sawmill. On him depends the quality and grade of lumber produced from each log. Very few logs are perfect either in the matter of shape or in freedom from customary defects, such as rot, punk, knots and checks, and the profit of an operation very often depends upon whether the sawyer gets all of the available good lumber out of the logs to the best advantage.

There are two ways, in general, of sawing lumber, namely, (1) flat grained and (2) quarter sawed or edge grained. The larger percentage of lumber is cut flat grained. Lumber used later for flooring, furniture, cabinet work, interior finish, etc., is often cut edge grained both because of its more pleasing appearance and because it will wear more smoothly and evenly and will not splinter. Edge grained lumber is that cut along the radius exposing the medullary rays which are large in such woods as oak, beech, etc. The cut is at right angles to the annual rings, with the resultant pleasing effect.

Seasoning. — After coming from the sawmill, lumber ordinarily is seasoned for from four months to a year, depending on the species, before being placed on the market. This reduces the moisture content and therefore the weight so that the cost of shipping is materially reduced. Every one knows how badly green lumber will shrink when used in a house or for general purposes. About one-half of the total weight of wood is made up of water.

Lumber is usually stored in regular and even but open piles in the yard so that the air will freely circulate through them and aid in drying out the wood. Hardwoods are heavier and more dense than conifers so that they require a longer time in which to season. Thick lumber also requires more time to season than thin lumber.

Methods of artificially seasoning lumber by means of the dry kiln and steaming processes have been in use for a long time. Only the best grades are usually seasoned in this way however. A dry kiln is an enclosure in which lumber is rolled on trucks and heated up to a sufficient degree to expel the moisture. If done too rapidly and in too dry a heat, the lumber will check and warp and split. Steaming lumber has also been successfully used to hasten the process of seasoning.

Planing Mill. — In connection with most of our sawmills, planing mills are run to plane flooring, interior finish and the better grades of lumber, later used for furniture, cabinet work, doors and sash. Planing is also done to save on freight rates. It is said that the cost of planing only one side of lumber will in many cases be justified in the lower freight rates that can be obtained as a result of this decreased weight.

Lumber is always seasoned before planing because planed lumber will split and check more easily than the roughly sawn material. This is due to the greater exposure of pores at the surface in the case of rough sawed lumber.

Typical Operation.

The methods of logging in this country vary widely in different forest regions. They vary with (1) the size of the logs, (2) character of topography, (3) climate, whether snow logging is permissible or not, (4) development of the region and (5) labor. Heavy logging machinery is required with the big Douglas fir and redwood logs of the Pacific Coast, whereas horses and mules are usually used in most regions. The sawmill operations are very similar all over the country and vary only with the size of the mill.

A typical American logging operation may be found in the northern forest where the chief trees to be cut are the white and red pine, beech, birch, maple and basswood, with some hemlock and white cedar in the swamps.

One or several rough log camps holding about 60 men each are constructed and haul roads are built before the snow comes. By the first of January there will be sufficient snow on which to haul the logs to the railroad landing and by this time there will be large quantities of logs waiting on skidways in the woods to be hauled out. Sleighs are then brought into service and the logs hauled down to the landing. About 3000 to 10,000 board feet are hauled at each load. The logging crew continues to fell and " buck up " the trees until the snow melts when the work in the woods will stop. Whenever a drivable stream is available, the pine and hemlock logs are sent down to the mill by this method. But the heavier hardwoods require a direct haul by team or the railroad.

The cost of a typical logging operation of this kind, based on charges per thousand board feet, would be about as follows: This includes the cost of logs from stump to the mill.

Felling and bucking	$0.80
Swamping and road building	0.40
Skidding	1.20
Hauling to landing	1.50
Railroad (construction and maintenance)	1.30
Camps (toting, building, repairs, blacksmith shop, etc.)	0.20
Superintendency — overhead charges	0.40
	$5.80

In figuring the net profit on the operation, stumpage charges are usually added to this cost. Woodsmen usually get about

FIG. 36. — A PORTABLE SAWMILL IN THE SOUTHERN APPALACHIAN MTS.

This type of mill is used mostly in connection with woodlots and small tracts of timber. A complete mill sawing from 3000 to 8000 board feet per day can be set up for between $1000 and $1500.

$30 per month and board. The cost of their board is charged against the particular work on which they are employed. Tools are furnished by the employer. In an ordinary camp of 60 men there will be about 35 horses used in skidding and hauling the logs.

The mill is located on some railroad, most convenient to the logging operation, and where there will be water for log storage and sufficient room to lay out the lumber yards. The sawmill operation can best be described by following a log through the mill. As it comes up the "jacker chain" from the pond, the scaler at the entrance to the mill scales the log and pulls a lever which kicks the logs down the log deck. When needed, the sawyer operates the "log stop and loader" which releases the lowest log and holds back the others. Then by use of a "steam nigger" he adjusts the log on the carriage. Two men usually ride on the carriage to keep the log in place. The carriage carries the log against the saw and the boards fall off on the live rolls which carry them to the edger which cuts off the uneven edges and the bark. The lumber next goes ahead to the trimmer which squares off the ends and as it comes out, it is graded and sent out into the proper pile in the yards.

All waste from the various saws is ground up for fuel to be used under the boilers or is sent to the waste pile. The slabs and defective logs are worked up into lath, box board and small stock.

The cost of manufacturing may be summed up as follows on the basis of a thousand board feet:

Log pond	$0.15
Sawing	2.50
Sorting and piling in yard	1.00
Planing	0.75
Dry kiln	0.50
Loading on cars	0.50
Insurance, repairs and depreciation	0.50
	$5.90

Portable Sawmills.

With the rapid cutting of our larger forest tracts, there are small bodies of timber scattered about, in woodlot holdings and second growth timber, that do not justify putting in a large logging operation.

Small portable mills are used to cut up these small tracts. They can be moved about from one place to another and are used especially to supply the local demand for lumber. They are occasionally used to saw up ties and mine timbers. The whole equipment only costs from $1000 to $1500 and can be run with from 4 to 6 men. They usually turn out only from 3000 to 10,000 board feet in a day. In Virginia and North Carolina there are over 2000 sawmills in each state. Most of these are of the small stationary or portable variety. Sometimes portable mills are run to turn out a special product, such as material for spools and bobbins, wood turning, furniture or cooperage stock, railroad ties, etc.

Very commonly a portable mill owner will estimate the timber on a small tract containing 100,000 or a 1,000,000 board feet and pay a lump sum for the standing timber alone. Most of our woodlot timber is worth from $3 to $10 per thousand board feet on the stump. Our best white oak, walnut, hickory, ash and white pine often bring from $10 to $16 per thousand as they stand in the woodlot.

As our virgin stands become more and more depleted, the small and portable sawmill will come more into greater prominence. In the future the large sawmill cutting several hundred thousand feet of lumber per day will be a thing of the past. Much of the timber from our National Forests, especially in the Rocky Mountain region, is now being sold to small mills to meet the local demands. In the Central States and East many farmers have portable mills that can be profitably used during the winter when the other work about the farm is not pressing.

Forestry and Lumbering.

Since about three-fourths of all the standing timber is now in the hands of lumbermen and private interests, it is a matter of great concern to have them adopt at least conservative measures to protect and perpetuate the timber supply. Lumbermen naturally look upon the work of the woods as a purely commercial or business proposition. They are only interested in adopting those measures which the forester can point out to them in a way that will better their profits.

The measures that the lumbermen are now coming to adopt, and in which the trained forester can help him are as follows:

(1) *Closer utilization of the raw products of the forests.* The first conservative measure adopted has been the cutting of lower stumps. Formerly from 10 to 15 per cent of the best part of the trees have been wasted by cutting stumps from 2 to 6 feet in height. In deep snow in winter much waste has been occasioned by cutting unnecessarily high stumps. Logs are also being taken down to a smaller diameter limit in the tops. On some operations where a diameter limit of 8 inches has been used it has been shown that by cutting to 5 or 6 inches much good timber can be saved. Partially defective logs are also being used for various bi-products of the forest if not for lumber. Care in sawing the logs in proper lengths, and in avoiding crooks, etc., has also resulted in a great saving. In the past much good timber has been used for skids, building camps, corduroys, etc., when inferior woods could just as well have been used.

(2) *Fire protection.* The lumberman is beginning to realize that a little money spent on protective measures is good insurance on standing timber. By the annual expenditure of a few cents per acre in the form of patrol, connecting telephones

and trails with lookout points, using spark arresters on all engines and in some cases disposing of all brush and slash after logging by piling and burning it, much valuable timber can be saved for the future.

(3) *Careful surveys and timber estimates.* It has been demonstrated that the expense of a careful topographic survey and estimate by species has been more than justified in reduced cost of railroad location, systematic logging and knowledge of the exact amount of available timber. Besides this a forester can be of considerable service in general surveying, locating haul roads, railroads, cruising of all kinds, etc.

(4) *Systematic organization of the whole operation.* It is a notable fact that in spite of the great size and variety of the lumber industry and the fact that it employs more men than any other industry, no systematic or scientific management has been worked out to advantage. Foresters, after sufficient training, are becoming of great help in planning and executing the work of large operations, devising means of economy and in providing for the profitable utilization of bi-products of both the mill and the woods that would otherwise be wasted.

Many operations that are sufficiently large are providing for a second cut. Foresters can determine what young trees are to be left to advantage, how to secure natural reproduction and how to protect the young growth. Several lumber and pulp companies are now reforesting large areas of land denuded by both the axe and fire. In Massachusetts several box board companies are planting white pine intending to cut it at an age of from twenty to forty years.

Future of the Industry.

Inasmuch as the total remaining stand of timber is estimated at 2,500,000,000,000 board feet, it is only a question of a comparatively few years, with our ever increasing consumption of lumber, before a definite shortage in our supply will be reached. We are now using twice the amount per capita that we did approximately fifty years ago and we are also paying now about twice the amount of money for our forest products that we did only about thirty years ago.

It can readily be seen that the products of the lumber industry will cost more and more in the future and that they will play an important part in our national economy. This means that the lumber industry will gradually adopt more and more conservative measures in cutting off the forests, and in time many of them will be managed on a continuous yield basis. Foresters will be employed to carry out these measures in the same way as we find in the older countries of Europe.

However, it may be said that it is the duty of the state and national governments to practice forestry on a more intensive basis than private interests and there is no doubt but that much of our forested lands, now held by lumbermen and others, will be purchased or taken over by the various states and the federal government. Our state holdings are now increasing very rapidly and the government has spent millions of dollars for the purchase of National Forests in the Southern Appalachians and the White Mountains.

BIBLIOGRAPHY.

Bryant, R. C. Logging. John Wiley and Sons, New York City.

Cary, Austin. Practical Forestry on a Spruce Tract in Maine. U. S. Forest Service Circular 131.

Defebaugh, E. H. History of the Lumber Industry in America. American Lumberman, Chicago.

FOLEY, JOHN. Conservative Logging at Sewanee, Tennessee. U. S. Forest Service Bulletin 39.

FOX, W. F. A History of the Lumber History of New York. U. S. Forest Service Bulletin 34.

SACKETT, H. S. Past and Present Prices of Forest Products. Vol. 2, Report of the National Conservation Commission, Senate Doc. 676, 2nd session 60 Congress.

SCHENCK, C. A. Logging and Lumbering or Forest Utilization. Darmstadt, Germany.

WILLIAMS, ASA S. Logging by Steam. Forestry Quarterly. Vol. VI, No. 1, pp. 1–33.

CHAPTER X.

WOOD UTILIZATION.

General.

Wood has always been an indispensable commodity in our civilization and, as one of the renewable natural resources, it will always continue to be. It is the most widely used commodity, outside of food and clothing, supplying the need not only for fuel purposes but for construction material and the various arts and industries that use wood in one form or another.

Our forests supply various medicines, naval stores, paper, maple sugar and syrup, wood tannins, oils and chemicals, as well as lumber.

In former years only the best trees were felled and utilized. The finest white pine and oak trees were commonly used for purposes to which inferior trees were just as well suited. We are now using vast quantities of hemlock, red gum, beech, tupelo and others that were not considered worth logging twenty years ago. More hemlock is now being cut than any other species in Pennsylvania, Michigan, Wisconsin and New York.

In a few more years, we will be forced to utilize second growth and the inferior species exclusively. Each species will then be put to the use to which it is best adapted.

One of the greatest problems that we are facing in our forestry practice in this country is the closer and more complete utilization of the raw products of the forest. Besides the enormous quantities of wood destroyed by forest fires, insects and fungi, we actually use only about 40 per cent of the total volume of the trees felled in our lumbering opera-

FIG. 37. — TYPICAL LOG STORAGE POND AT A LARGE SAWMILL IN OREGON.

We use annually about 40,000,000,000 board feet of lumber besides large quantities of ties, poles, posts, fuel-wood, mine timbers, and wood for pulp, veneers, excelsior, etc.

tions. The remainder is generally wasted in the following ways:

	Per cent.
Loss of wood in the stumps	10
Loss of wood in the tops and branches	14
Waste at the sawmill:	
Sawdust	11
Bark	8
Slabs	10
Edgings and trimmings	7
Total	60

In many cases, this waste is being cut down by using sawdust for fuel, slabs for laths, box boards and small wooden articles, and general waste for making pulp, or wood distillation products. Lumbermen are also cutting the trees closer in the woods. But the general public is as much to blame as the lumbermen, for lumber is wanted only in even foot lengths and even inches in width, short pieces are not in demand and the railroads refuse to offer a sufficiently low freight rate so that the low grade lumber can be marketed at a profit, or in some cases, even near the cost of manufacture.

We are securing very complete utilization in some wood using industries, such as the wood pulp, excelsior, wood distillation, veneer, furniture, and wood novelty industries, but there is a correspondingly serious waste in the tight cooperage, turpentine and lumber industries and in the hewing of cross ties.

With the inevitable increase in the market value of our forest products, and the improvement in the transportation facilities, our annual wood waste will be reduced to a minimum as it now is in European countries.

Wood Production.

The production of forest products, especially lumber, has been on the steady increase within the past few decades. For instance, in 1880, the lumber production amounted to eighteen billion board feet. At the present time it is about forty billion board feet.

The annual supply of wood products can best be shown in the following table. This is exclusive of the minor forest industries which are described later in this chapter.

Annual production of wood, exclusive of the amounts used in the so-called minor forest industries:

Class.	Amount.	Cubic feet.
Lumber	40 billion board feet	9,000,000,000
Firewood	90 million cords	7,000,000,000
Cross ties	150 million pieces	1,500,000,000
Poles	3800 thousand pieces	1,600,000,000
Posts	500 million pieces	
Round mine timbers	(Various sizes)	200,000,000
	Total	19,300,000,000

Altogether about 20,000,000,000 cubic feet of wood are consumed annually in this country. The principal uses of the more important species will be found in the appendix.

Wood Consumption.

Use of the Lumber Cut. — The above table in a general way shows how our wood is consumed as well as produced. The principal item, lumber, however, is further used and manufactured in a great variety of ways. Much of it is used in the rough for general lumber and construction purposes, but the better grades are planed and used later for flooring, interior finish, furniture, vehicles, car construction, sash and doors, woodenware, implements and a long list of similar lines of usage.

An investigation of the wood using industries of about twenty representative states shows that the lumber cut is used in the following different ways, together with the percentage of each:

	Per cent.
Planing mill products, such as sash, doors, flooring and general mill work	30
Rough lumber and structural timbers	26
Boxes and crating	10
Car construction	5
Furniture	3
Vehicles	2
Agricultural implements	1
Musical instruments and woodenware	1
Total	78

The remaining 22 per cent are made up of export lumber (7 per cent) and miscellaneous wood uses such as boat and ship building, trunks, all kinds of handles, wooden fixtures and appliances, etc.

American versus European Conditions. — This country is the greatest wood consuming nation on the globe. We use 230 cubic feet per capita as against 200 for Canada, 37 for Germany and 25 for France. Moreover, Canada has 60 acres of forest per capita whereas we have only 6.

But the greatest contrast is with European conditions. As opposed to our use of only about 40 per cent of the trees felled in our forests, it is estimated that in Germany, about 96 per cent of all the wood that is grown are utilized. Everything, literally, is used in one way or another. Even the stumps are grubbed out and the smallest twigs are utilized. There is no waste due to destructive forest fires and very little insect or fungous damage.

The reason for the close utilization in Germany is the scarcity, and therefore the high value of wood supplies, together with their national spirit of economy, good transportation facilities and demand for all classes of wood products.

We are rapidly approaching the same situation in this country in respect to wood utilization. Our better species are rapidly disappearing and we are already using inferior trees and those that are defective and knotty. This country was originally endowed with a greater variety and better quality of trees than any other nation. We have been wasteful and prodigal in their use. In Germany, there are only about five important species that are adapted for forestry purposes. In this country we have at least twenty-five species that are correspondingly valuable to suit our conditions.

Wood Substitutes. — It is often said that when our timber resources are depleted, we will find satisfactory substitutes. But in spite of the introduction of brick, steel, concrete and other structural materials, we find that our per capita consumption of lumber to-day is twice as much as it was fifty years ago. Substitutes are welcomed in the forestry situation but it is extremely doubtful if wood will ever cease to be a commodity of wide usefulness.

It is only a natural sequence of the old economic law of supply and demand, that in the future, brick, cement, structural iron and other materials will take the place, to some extent, of wood. However, in the foregoing list wood used for general lumber and structural timber purposes comprised only 26 per cent of the whole lumber output. The remainder is used in an indefinite variety of ways, for which it is extremely doubtful if substitutes will displace wood.

In Germany substitutes for wood have been introduced wherever practical, and one finds nearly all the houses constructed, on the exterior at least, of stone, brick, etc. Steel has taken the place of wood for office furniture, bedsteads and bridges to a large extent.

Even in this country, we have concrete posts and poles, and to some extent, steel cars, cement floors, etc., and as other materials become cheaper and wood more expensive, lumber will be somewhat displaced.

But with the introduction of wood substitutes we are constantly finding new uses for wood. For instance, it has been determined that, on account of its sanitary features, durability and freedom from noise, wood block paving is superior to other forms of street paving. In the wood pulp industry the demands for wood are increasing very rapidly and new species are being introduced to supply the demand.

The attempted use of steel and concrete ties has not been

a success, due to the lack of resilience and elasticity peculiar to wood, that is, the steel or concrete ties are too hard and rigid. The steel ties corrode rather easily and both the steel and concrete ties involve a large initial expense.

Minor Forest Industries.

The following are classed as the minor forest industries:

Wood Pulp. — Paper has been made from woody fibers ever since the papyrus was used for this purpose in ancient Egypt. But it is only within comparatively recent times that wood has been used on a large scale. Spruce now supplies about 60 per cent of all the wood used for paper. It is especially suited to this purpose because of its comparative freedom from resin, the length and strength of its fibers, and its softness and availability.

Our spruce timber, however, is being rapidly cut off and many others are being introduced as follows: hemlock, poplar, balsam fir, white and yellow pines, beech, maple, white fir and the western spruces. Practically every important timber tree is now used, at least to some extent, in paper manufacture. The hardwoods and resinous conifers, however, only make the rougher and coarser grades of paper.

About 4,300,000 cords of pulp wood are consumed annually. Canada now supplies over one-half of our spruce for this purpose. The leading states in the industry in order are: New York, Maine, Wisconsin, New Hampshire and Pennsylvania. New York consumes over 1,000,000 cords annually. About $9 to $12 per cord are now being paid at the mills for spruce pulp wood.

There are, in general, four methods of making pulp, as follows:

1. The mechanical or ground pulp process: This is a mechanical process in which the wood fibers are ground up

into pulp by pressing the bolts of wood against a revolving stone. The bark is always rossed off first in all methods. Spruce is the principal wood reduced by this process.

2. The sulphite process consists of boiling small chips of wood in a large cylindrical digester, containing calcium sulphite, for about sixty hours. Hemlock is reduced principally by this process.

3. The soda process consists of boiling the chips in a digester, containing caustic soda, and bleaching with chlorate of lime. Practically all of the poplar is reduced by this method.

4. The sulphate process can be used with pine, spruce or larch and consists of using sodium sulphate as a chemical and cooking in a digester as above.

The principle involved in all methods is to grind or chip the pulp wood and reduce the wood fibers to a fluid pulp. Ground pulp is less interlaceable than the chemical pulp and contains lignin as well as cellulose. When paper made from mechanical pulp is exposed to the light, it turns a brownish or yellow color. Newspaper is composed of about 75 per cent ground pulp and 25 per cent of sulphite pulp. The best grades of book paper usually contain no ground pulp at all.

By whatever process the pulp is made, it is washed and screened carefully and pressed into thick sheets which then go to the paper mill.

In the paper mill the pulp is beaten up into a semi-fluid mass; resin, clay and other ingredients are added to give it body and added strength, and coloring matter is mixed in if desired. After being washed and screened it is passed through a long line of presses which gradually eliminate the moisture and dry out the paper.

The manufacture of paper involves the investment of enormous sums of money for power development and machinery, much of which is highly specialized.

The paper industry utilizes the raw product of the forest more completely than any of our wood-using industries. It also offers a splendid field for the utilization of the bi-products of other industries. At the present time about 280,000 cords of mill waste, in the form of slabs, edgings and other refuse, of the large sawmill are now used for paper pulp. They are converted into pulp by the sulphite process which uses the wood in the form of small chips and can therefore utilize mill waste to advantage.

One cord of wood yields roughly about one ton of air-dried ground pulp or about one-half ton of chemical pulp. Some pulp mills consume as high as 200 cords of wood in a day.

Wood Tannins. — The use of tannin, extracted from the bark and wood of certain species, in the manufacture of leather, by rendering the skins more durable, pliable and supple, has been in practice for a long time. In fact it is one of the oldest known industries. For sole and other heavy leathers, wood tannins are preferred, although this source of material is becoming so rapidly depleted that chemical tannins are being substituted to some extent.

Wood tannins are mostly produced from hemlock bark, oak bark and chestnut wood. Over a million cords of bark worth $8 to $10 per cord and over 200,000 tons of chestnut wood worth about $30 per ton are used every year. Quebracho, gambier, mangrove bark, myrobalan nuts and other tropical sources of tannin are now being imported into this country in considerable quantities.

The following per cents of tannin are generally recognized as being contained in the following sources:

Species.	Material.	Per cent.
Eastern hemlock	Dressed bark	13.1
Western hemlock	Dressed bark	15.1
California tanbark oak	Dressed bark	16.1
Chestnut oak	Dressed bark	6.2
Chestnut	Wood	8.0
Quebracho	Wood	24.0

Hemlock and oak bark can only be peeled from the trees felled after the sap rises in the spring. Oak is best peeled in May, whereas hemlock can be easily peeled from May to September. Before hemlock lumber could be marketed profitably, enormous quantities of this species were felled in our eastern forests for the bark alone. As the bark is peeled, it is seasoned in the woods and shipped and sold by the cord, weighing about 2240 pounds.

The average cost of marketing hemlock bark per cord is about as follows:

Peeling and piling	$2.50
Hauling to the railroad and loading	1.50
Railroad transportation	0.30
Supervision	0.35
Total cost per cord	$4.65

A cord of hemlock bark usually brings between $6.50 and $7.50 delivered on the markets.

An ordinary acre of hemlock forest running about 10,500 board feet to the acre yields about 7 cords of bark. One cord of bark can therefore be secured from about 1500 board feet of standing timber.

Naval Stores. — This industry covers the production of commercial spirits of turpentine and rosin by the distillation of the resinous exudation of certain species of pine, chiefly longleaf pine of the South.

About 29,000,000 gallons of turpentine and 3,200,000 barrels of rosin, valued at over $25,000,000, are produced annually. In many sections of the South, especially in Florida, it is the principal industry. After being tapped for three or four years under the present wasteful methods, the trees are logged. Tapping for turpentine does not injure the quality of the wood for lumber or other wood uses.

FIG. 38. — A TURPENTINE STILL ON THE FLORIDA NATIONAL FOREST.

They are used to distill the crude resin, collected from the longleaf pine, into turpentine. The residue after distillation is the rosin of commerce.

To obtain the raw resin, the trees are first " boxed," which consists of cutting a collecting box into which the resin drains in the base of the trees. During the warm season, when the sap flows freely, at regular periods the bark is chipped in V-shaped strips so that the flow is continuous. This consecutive chipping is carried up the tree as far as a man can reach with his " hack " and leaves the white face so common all over the South. There are sometimes three or four faces on a single tree, depending on its size.

The crude resin is collected six or eight times during the

season and sent to the still. Much of the resin does not flow down to the box and has to be scraped off.

The still consists of a large copper kettle under which is a large fire box and a copper condenser or worm about 175 feet long. About ten barrels of crude resin are emptied into the kettle and after a fire is started it is boiled for three or four hours. The distillate from this process, after passing through the condenser, is called turpentine, a light, watery substance, which is barreled and shipped to the markets to be used for paints, varnishes, coloring and a great variety of chemicals and medicines.

After the boiling process is finished, the residue in the kettle is carefully screened and poured out into barrels. This is the rosin which is graded according to its clarity and whiteness, and is used for glue, varnish, soap, soldering and the manufacture of sealing wax, etc.

Dr. C. H. Herty has devised a new method in place of the old and wasteful way of tapping the trees which are so easily injured by fire getting into the boxes and being either burned completely or later thrown by the wind. Instead of cutting a box at the base of the tree, two incisions are made, into which are inserted galvanized iron gutters. These direct the flow of sap into earthen or metal cups hung on zinc nails. The chipping is carried up the tree as in the old method. Several variations of the Herty method have also been introduced.

The advantages of the Herty or cup and gutter system are as follows:

1. The yield is greater and better because the cups can be moved up the tree and very little is wasted in collecting. Less dirt and bark is gathered in using this method.

2. It does not weaken the tree physically so there is less danger from wind throw.

3. The danger from serious damage by fire is much less.

Many operators in the South are now adopting this method. The only objection to it is the initial expense of purchasing and installing the cups and gutters. In the long run, however, it has been demonstrated to be much more efficient and economical.

In Europe, Norway spruce and maritime pine are the native sources of naval store products. The latter is being introduced in this country with the view of perpetuating our vanishing supply.

Cooperage. — Cooperage refers to the use of wood in the manufacture of receptacles and containers, such as barrels, kegs, casks and tubs. Tight cooperage is that kind used to hold or ship liquids, especially oil and tierce, wine and liquors, pork, etc. Up to the present time white oak is practically the only species used for tight cooperage because of its impermeability and the fact that it does not impart any disagreeable odor or flavor to the contents. White oak, however, is becoming so scarce and expensive that other woods, such as red oak, red gum and cypress, are being introduced into the industry especially for heading.

Slack cooperage is used largely for marketing agricultural products, such as vegetables and fruit, and for shipping flour, sugar, cement, crockery and a great variety of similar articles. Availability and cheapness determine more than anything else the kinds of woods to be used for slack cooperage. However, woods which dry quickly, steam well and which do not contain oils, resin or other substances which may injure the contents are especially in demand for certain uses. The veneer barrel is coming rapidly into use.

The following are the principal woods used for slack cooperage staves: red gum, pine, beech, elm and chestnut. The following are used in order for heading: pine, red gum, beech, maple and oak. Elm is the only wood used for hoops. Several machines are used to manufacture the staves and heading which are often made in separate plants. They are then shipped and assembled where first placed in use.

Slack cooperage offers a splendid field for the close utilization of forest products. Limbs, tops and defective logs are often used, and in many cases after logging is finished on a lumbered area, much material is gathered for staves and heading.

There are about 1,300,000,000 staves, over 100,000,000 sets of heading and 350,000,000 hoops used annually for slack cooperage. One set of heading, 15 staves and 6 hoops are usually used in the average slack cooperage barrel.

Considerable waste is occasioned in the manufacture of tight cooperage, both in the woods and at the finishing plant. Only the best white oak is used. The staves are usually sawed, but many are also "bucked and split" in the woods or are hewed. Arkansas, Tennessee and Kentucky are the center of the industry. Enormous quantities of white oak are shipped to European points, especially France. About 350,000,000 staves and 30,000,000 sets of heading are produced for tight cooperage annually.

Wood Distillation. — This industry comprises the utilization of certain products of the forest in the distillation of wood either by the destructive or steam process. The principal products of the industry are wood alcohol, acetate of lime, turpentine, tar and charcoal, and various chemical distillates such as acetone, formaldehyde, wood ashes, pyroligneous acid, pyrolignite of iron and wood creosote.

In the North where the industry has been mostly developed, the hardwoods, principally beech, birch and maple, are used and the products are chiefly wood alcohol, acetate of lime, and charcoal. Over 1,550,000 cords of hardwoods are used annually and the principal states engaged in the industry are Michigan, New York and Pennsylvania.

In the South the industry is confined to the yellow pines, the principal products being turpentine, tar and creosote. Over 190,000 cords of softwoods are used annually and the industry is on the rapid increase especially in the South and in the Douglas fir region of the northwest.

FIG. 39. — RETORTS USED IN THE WOOD DISTILLATION INDUSTRY.

The wood is placed in these air-tight ovens which are heated to high temperatures. The gases that pass off are collected and used as the basis of several wood chemicals. The residue, charcoal, was formerly made in the wasteful open air pits.

Large quantities of limb wood, stumps and mill waste, including sawdust, slabs and defective material, are now being used in this industry so it is playing an important part in the closer and more effective utilization of the raw products of the forest.

The process consists largely of the dry distillation of wood, differing from the old method of charcoal burning in that the resultant gases that pass off are utilized instead of being allowed to escape in the air. For this purpose, the wood, after being thoroughly air-dried for about a year, is placed in long cylinders called retorts, usually about 30 to 40 feet long and 5 to 6 feet in height or diameter. The doors are clamped air-tight and the cylinders are heated to a high temperature. The gases that pass off in this heating process are led through condensers and utilized. The non-condensing gases are used for fuel in the boiler room. With conifers, tar forms at the bottom of the cylinders and is collected through a system of pipes. The condensible gases are further distilled into the products mentioned above.

One cord of beech will yield about 190 pounds of acetate of lime, 9 gallons of wood alcohol, 14 gallons of tar and about 1000 pounds of charcoal.

One cord of highly resinous wood, such as longleaf pine, will produce about 24 gallons of turpentine, 33 gallons of pyroligneous acid, 120 gallons of tarry and oil products and 56 bushels of charcoal.

Veneers. — Although not a minor forest industry in the same sense as the others, the manufacture of veneers has taken rapid strides during recent years and the industry has been of great service in effecting a more complete utilization of our forest products. A veneer plant is frequently maintained in connection with a large lumber operation in the same manner as the naval stores, wood tannin, excelsior, and other industries.

Veneers are thin slices or sheets of wood. Formerly they were largely made of the fancy and more valuable woods, such as mahogany and walnut, to cover less valuable woods in the manufacture of furniture, cabinet work, etc. They were known and used even in old Roman times. At the present time, however, the demands for veneers are so great that cheapness and availability of supply determine the species that are used.

They are used not only for all kinds of furniture and cabinet work, but for baskets and crates, cooperage, trunks, doors, musical instruments and many other uses where thin sheets of wood are required. Doors and furniture made of "built up" stock, that is, several layers of veneers with the grain running cross-ways, are stronger and less apt to crack and warp.

Veneers may be cut down to a thickness of $\frac{1}{100}$ of an inch but those less than $\frac{1}{40}$ of an inch are seldom used. They are principally cut in thickness of $\frac{1}{24}$ of an inch. For cooperage or other purposes of similar nature, they are cut to a thickness of $\frac{1}{4}$ inch or more.

There are three different methods used in making veneers: (1) Rotary cut, (2) Sliced and (3) Sawed veneers.

Most of our veneers are rotary cut. With this method, a log is generally boiled for several hours in water, then set in a lathe and turned against a sharp knife. A continuous sheet is peeled off and cut up into desired sizes. The logs can be cut down to a diameter of about six inches. This core is then sawed up for box boards or other material.

For slicing veneers, a special machine is provided to slice off sheets of any desired thickness either radially or tangentially. The knife is usually stationary, while the log or timber is sent against it in a vertical movement. At each shearing stroke the log is moved up the desired width.

Sawed veneers are usually cut on a circular or band saw with a very narrow kerf, so that there will be the minimum amount of waste in sawdust. The most expensive veneers are made by this process.

About 500,000,000 board feet of wood are used annually

for veneers in this country. The following are the principal woods used in order:

Species.	Per cent.
Red gum	33
Yellow pine	8
Maple	8
Yellow poplar	7
Cottonwood	7
White oak	6
Birch	5
All others	26
Total	100

Altogether a great variety of species are used for veneers. Mahogany, Circassian walnut and Spanish cedar are the principal imported woods used.

Excelsior. — Excelsior is used for mattresses, upholstering, shipping a great variety of articles, such as glass-ware, furniture, crockery, metals, etc., for filtering purposes and to some extent for mattings and rugs. Wood wool is an especially fine grade of excelsior.

The production of excelsior is on the rapid increase. Some of our large stores use as much as 35 tons per month. One toy company uses about 30 bales of 120 pounds each per day.

Basswood makes the best excelsior because its wood is springy and soft and, when thoroughly dry, does not lose its elasticity. Cottonwood, poplar, white pine and buckeye are also used. The relative scarcity of these woods has caused the use, in addition, of yellow pine, yellow poplar, birch, red gum, spruce and maple.

There is practically no waste in this industry. The peeled bolts are placed in frames which send the wood against sharp spurs or knives. These shred off the excelsior. This is gathered and packed in bales weighing from 100 to 125 pounds apiece. One standard cord of basswood will yield about 1500 pounds of excelsior.

Maple Syrup and Sugar. — All of the species of the maple family produce a sap that can be reduced to maple syrup or sugar, but only one, the sugar or hard maple (Acer saccharum), is of commercial importance.

The practice was first learned from the Indians. They employed, however, very rough methods of both obtaining the sap from the tree and in boiling it down to a sugary substance.

Although the sugar maple grows in practically every state east of the Mississippi River, best results for sugar and syrup seem to be secured in the northeastern states where the cold nights alternating with warm sunny days in the early spring seem to be conducive to the greatest flow of sap. Trees are first tapped about the middle of March and the flow continues to about April 15, depending upon the region. A hole about three-eighths inch is bored with an augur to a depth of about two inches and sloping to carry out the sap. A wooden or iron spiggot or spile is inserted in the hole which conducts the sap to a bucket hung on the spiggot or suspended from a nail. Two holes are usually bored in each tree and new holes are made each year, the old ones usually healing over. The sap is gathered from time to time and taken to the sugar house. Formerly the sap was boiled down in a large kettle over an open fire in the woods. In connection with the large sugar orchards, large evaporators are now maintained with the heat often furnished through steam pipes. This method results in a greater and much cleaner product.

The "sugar bush" is an important part of nearly every woodlot in the maple sugar region, namely Vermont, New Hampshire, Northern New York and Northern Ohio, and considerable revenue is obtained from this source. The industry has been discredited to a considerable extent in some places by the introduction of substitute flavoring extracts and of diluting materials. Vermont protects the industry through a system of state inspection.

A good healthy tree will often produce 25 gallons of sap per season. It requires about 12 gallons to make two pounds of sugar or one quart of syrup. The sugar is made from the syrup by simply boiling it down to a consistency of wax, when it will crystallize after being poured into forms.

Most of our sugar orchards are maintained in poor condition, in that cattle are allowed to graze in them, preventing a renewal of productive trees from the young stock. Other species should also be removed to give all the growing space to sugar maples. Too many or too large holes should not be bored in the trees to get the best results. The tapping of the trees for sap, if done conservatively, does not seem to injure the health or growing power of the trees.

Miscellaneous. — Besides the above wood using industries, there are a great number of less importance. Among them may be mentioned the furniture, box board and wooden novelty industries, the spool, shuttle and bobbin works, the vehicle and implement industries, the toy industry, etc., all of which use large quantities of wood. For instance, in Wisconsin, 25,000,000 board feet of basswood alone are consumed annually for just woodenware. These, however, are purely industrial enterprises and are merely a matter of general interest to the forestry profession.

Methods of Securing Closer Utilization.

In the Woods. — American lumbermen are rapidly learning and appreciating the benefits to be derived from the closer utilization of the raw products of the forest.

The first step taken was to lower the stump cut. The Forest Service requires a maximum height of sixteen inches and in some cases twelve inches on the National Forests. Sometimes the rule of cutting the stumps at a height equal to the diameter breast high is followed. On the Pacific Coast stumps were formerly cut from a spring board at a height of from six to twenty or more feet, thereby losing the best timber in the tree.

Lumbermen are also cutting closer in the tops. Formerly, logs were cut to a minimum diameter of eight or twelve inches in the tops. Now white pine is commonly cut down to five inches in the tops and practically all species in the East are cut to a six inch top diameter.

Skidways, corduroys, roads and camps formerly made from the most available trees are now being constructed of tops, limbs and inferior species. A great saving is possible on some operations in this respect alone.

Special industries for the use of tops, branches, defective logs, etc., are being introduced in connection with large lumber operations to utilize all of the available material, such, for example, as wood distillation and slack cooperage plants to use short pieces, and material down to six inches in diameter; wood pulp plants to use both woods and mill waste and charcoal pits.

At the Mill. — A great economy at the mill in saving former wood waste is in the use of improved machinery. The band saw, especially, has effected a great saving in sawdust alone since it cuts a kerf of about one-eighth of an inch, as against a kerf of three-eighths of an inch to one-half inch for the old circular saws. Improved re-saws, edgers and trimmers have also helped in turning out both more and better lumber.

However, the greatest saving at the mill has been the utilization of the slabs, edgings and trimmings that were formerly sent to the burner and wasted. Most of our large sawmills already use their slabs for the manufacture of lath from softwood stock. Hardwood slabs are gradually being utilized for the turning out of small squares and short pieces to be used for furniture stock, brush backs, handles, wooden ware, novelties, and toys, etc.

Some of our waste from chestnut lumber is being utilized for tanning extract. Waste from coniferous mills is being used for making wood pulp. In the year 1912, 280,000 cords of mill waste were used in this way.

Other mill waste is being used for wood distillation. Even the sawdust not required for the boilers to supply power to the mill, is sometimes shipped for use in ice packing, shipping brittle wares and stable bedding. Mills near populated sections are selling their mill waste to advantage for fuel purposes.

Considerable economy is also being brought about by improved methods of seasoning lumber liable to warp and check, such as dry kilns and steaming processes.

BIBLIOGRAPHY

Brown, Nelson C. Utilization at the Menominee Indian Mills, Neopit, Wisc. Forestry Quarterly No. 3, Vol. X, 1912.

Bureau of the Census, Washington, D. C. The following statistical circulars: —

Lumber, Lath and Shingles, Pulp-wood consumption, Slack cooperage, Tight cooperage, Wood distillation, Veneers, Wood tannins, Cross-ties and Poles.

Clapp, E. H. Conservative Logging. Report of National Conservation Commission, Senate Document 676, 60th Congress, 2nd Session, 1909. Pages 512–546.

Cline, McGarvey. The Lumbermen and the Wood Using Industries. Proc. Soc. of Am. Foresters, No. 2, Vol. VIII, 1913. Page 154.

Peters, J. G. Waste in Logging Southern Yellow Pine. Yearbook of U. S. Dept. of Agr., 1905. Pages 483–494.

CHAPTER XI.

WOOD TECHNOLOGY.

General.

Every species varies in its wood structure and it is this difference that we depend upon to identify our various woods.

A study of wood structure, together with the physical, mechanical and chemical properties, determines the uses to which each species is adapted. For example, we know that some trees, like chestnut, oak, and longleaf, are durable and are therefore used for railroad ties; Douglas fir is strong and stiff and makes a good construction timber; spruce has long strong fibers and is comparatively free from resin and so is used for paper pulp; mahogany seasons well and does not warp or twist and has a beautiful grain, all of which make it of especial value for fine furniture and cabinet work; elm is tough and therefore makes a good vehicle and hoop wood; white oak is impervious to liquids and does not impart a disagreeable odor or flavor and so makes a desirable tight cooperage wood.

Each wood has certain characteristics which distinguish it from others in color, weight, grain, strength, stiffness, etc. These characteristics determine the value of every wood for the various lines of utilization and therefore to a large extent, together with their available supply, their value on our timber markets.

Some of our species are becoming so rapidly depleted that substitutes are being introduced to take their places. In some cases, ingenious methods have been devised to imitate the grain and color of our more valuable woods. For example, red gum, a comparatively cheap wood, is largely used to imitate black walnut and Circassian walnut. In England it sells as satin walnut. Red birch is sold to a considerable extent as mahogany in furniture and cabinet work, and many new foreign species are being imported and sold as the true mahogany. A new method of graining cheap woods to imitate quartered white oak is now in common use.

The knowledge of wood structure is therefore vastly important in identifying the many different species that enter our American markets, both foreign and domestic species.

One of the matters of especial interest in wood technology is the chemical utilization of the enormous quantities of wood waste incident to our present methods of logging and lumber manufacture. New methods are being constantly devised to produce gases and distillates in commercial quantities from several of our timber trees. In Germany, cattle food has been made for some time from sawdust; new processes are being discovered in the chemical reduction of wood fibers for the manufacture of paper pulp, and improved methods of chemical distillation open up a large field for future development.

Wood Structure.

Wood structure and arrangement of the woody tissues may be said to underlie the principles that govern both the means of identification and the uses to which our woods are put.

Structure refers to the relative size, shape and form of the wood elements peculiar to each tree. It explains why some species are heavier, stronger and stiffer than others, why some are cross or straight grained, hard or soft, and why some have a tendency to check and split, while others season rapidly and without injury. The structure of the oak with its wide medullary rays explains why it has such a pleasing effect when this species is quarter sawed and polished to bring out the grain.

Bark. — The bark serves chiefly as a protective covering. It usually forms from 7 to 15 per cent of the whole contents

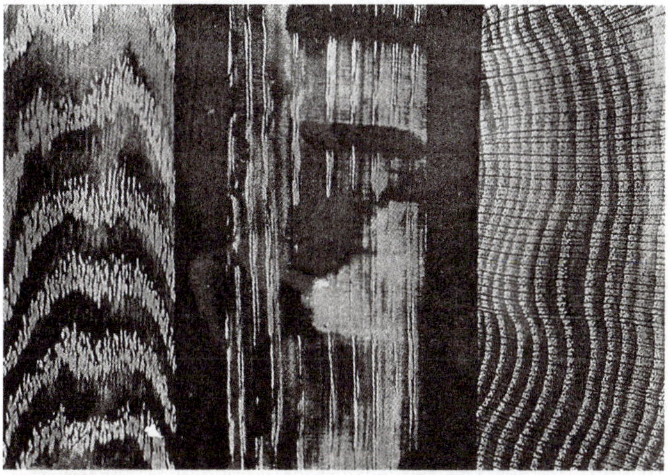

FIG. 40. — WOOD SECTIONS OF RED OAK (Quercus Rubra).

A typical ring porous wood. On the right is a transverse or cross section, in the center, a radial section, and on the left, a tangential section. Note the medullary rays and the spring and summer wood.

of the stem. It is much thinner on young than on old trees and is much thicker at the base than farther up the tree. There are successive layers of bark in the same way that successive layers of wood are laid on each year, but they are very indistinct.

The bark often serves as a means of identification. It is very thin in the spruces, cedars, birches, cypress, larch, longleaf pine, etc. It is usually thick in the case of Douglas fir, white pine, yellow poplar, redwood and some of the oaks.

The bark of many trees is of high value. For example, the bark of the cork oak (Quercus suber) of southern Europe supplies our cork of commerce; the bark of the hemlocks and most of the oaks is of high value for tanning purposes. Birch bark was formerly used for canoes; other barks are of high medicinal or fuel value. The fibrous inner bark of basswood is often used in the manufacture of fiber cloth, rough carpets and mattings.

Sapwood and Heartwood. — Inside the bark is the sapwood, usually light in color, varying in width from one-half inch to several inches. Within the sapwood and in the center of the tree is the heartwood, usually much darker in color.

The heartwood has practically no other function than that of mechanical support. On the other hand, the sapwood is a living part of the tree, serving largely to store up the starch and to conduct the unelaborated sap from the roots up to the leaves. At the extreme outer periphery of the sapwood is the cambium layer where the growth of the tree in diameter takes place by cell division during the growing season.

In young trees the percentage of sapwood is the greatest. It varies, however, with the species. The following species have thin sap as a rule: redwood, catalpa, locust, red cedar and yew. The following generally have wide sap: hickory, maple, ash, beech and some of the pines. In some species, such as cottonwood, willow, spruce, fir and hemlock, there is little difference in appearance between heart and sapwood.

The darker color of the heart is usually caused by the deposition of tannin, gums, resins, etc., and therefore the heartwood is heavier, more durable, and contains less moisture. These often render it the most valuable portion of the tree. In some cases, however, for handles, spokes, spools and certain other wooden articles, only sapwood is used as in the case of hickory for handle stock. Sapwood can be impregnated much more readily with chemical preservatives and in the natural state seasons with much difficulty. It is much more susceptible to decay, owing to its greater moisture content.

Annual Rings and Grain. — Grain refers to the direction and width of the growth rings. Sawed boards are often cross-grained because the line of the saw cut does not follow the grain. This explains why split wood is stronger than sawed wood. Therefore bolts for manufacture into spokes, handles, tight cooperage, etc., are preferably split rather than sawed.

Species with wide pith or medullary rays, such as oak, beech, and sycamore, yield a beautiful silver grain when sawed radially, or at right angles to the annual rings. Wavy and curly grain are common variations in the growth of many of our species, especially hard maple and yellow birch.

As explained in previous chapters, the width of the annual rings depends upon the species, conditions of growth, etc. All trees grow relatively fast in the spring and slowly toward the end of the growing season. The wood elements are therefore much coarser during the first part of each year's growth. This gives rise to the differentiation between spring and summer wood. This is brought out very markedly in the southern pines and in Douglas fir. The spring wood is light in weight and in color, whereas the summer wood is dark in color and heavy, giving the common ribbed appearance of these woods when cut across the grain. This can also be readily seen in any flat grained lumber of nearly every species.

Hardwood Versus Coniferous Wood. — Custom has been responsible for calling all our broadleaved trees, hardwoods, and all our evergreens, conifers. These, however, are misnomers in both interpretations because there are certain conifers, like larch and cypress, which are not evergreen since they are deciduous and drop their leaves in the fall. Then, too, many of our hardwoods are much softer in their wood structure than certain conifers or so-called softwoods and *vice versa.* For instance, softwoods, like larch, longleaf pine and Douglas fir, are much harder than so-called hardwoods, like basswood, willow, cottonwood, yellow poplar, buckeye and red gum. These terms are therefore merely for classification.

FIG. 41. — WOOD SECTIONS OF BIRDS-EYE MAPLE (Acer Saccharum). A typical diffuse porous wood. On the right is a cross section, in the center a radial section, and on the left a tangential section.

Another means of classification for the purposes of identification is according to the porous nature of the various woods.

On examination, in cross section, even under the microscope, it is found that the pores of the softwoods or conifers are not visible or conspicuous. Further classification of the genera and species is by means of the resin ducts, color, odor, taste, appearance of heartwood, etc. The hardwoods are divided into two broad groups: (1) Those that are ring porous, such as oak, ash, hickory, elm, chestnut, catalpa, and (2) Those that are diffuse porous, such as red gum, yellow poplar, willow, birch, maple, basswood and beech. The ring porous woods are those with a wide band of large pores in the spring wood. The diffuse porous woods are more homogeneous and continuous in their structure and texture.

Altogether about 80 per cent of our lumber cut is coniferous and it is used largely for construction and general lumber purposes. The hardwoods are used principally in specialized lines of utilization, such as furniture, cabinet work, vehicle and implement stock, and fine interior finish and flooring.

Seasoning.

General. — The percentage of moisture represented in the total weight of green wood is between 50 and 60 per cent. Most of our thoroughly seasoned lumber contains from 15 to 20 per cent of moisture. The immediate object of seasoning wood is principally to reduce the moisture content.

The reduction of the moisture content and therefore the ultimate objects of seasoning may be summarized as follows:

1. To decrease the danger from decay and therefore increase the durability of the wood.
2. To prevent warping, twisting, checking and shrinkage after the wood is placed in service.
3. To increase the strength and stiffness of wood.
4. To decrease its weight and therefore save on shipping charges.

When green, wood is very susceptible to fungous and insect attack.

Air seasoning is the usual method of decreasing the water content in wood. For the best grades of lumber, however, kiln drying or various steaming processes have been introduced to artificially hasten drying.

The rapidity of seasoning varies with:

1. The structure of the wood.
2. The size and shape of the wood.
3. The method of seasoning and piling, and the condition of the atmosphere.

Thus inch lumber dries four times as rapidly as four inch stock. White pine dries faster than oak on account of the difference in structure and density. Conifers usually require from two to five months for air seasoning whereas hardwoods require from six to ten months.

In dry kilns, temperatures of from 150 to 180 degrees Fahrenheit are usually used, and only a few days are ordinarily required for the drying process. Lumber, however, is sometimes air seasoned for a while before being dry kilned.

Sawed lumber dries more readily than planed lumber because of the greater areal surface exposed in the former case.

Lumber is usually piled in the direction of the prevailing wind so that the drafts of air will take out the moisture laden air in the alleys most readily.

Weight of Wood. — The various species vary a great deal in weight. Most woods float in water because they are lighter than the water which they displace. But this is so because of the air contained in the wood. When these air spaces are filled by water in the case of driving and rafting, the logs become " water logged " and sink. This is explained by the fact that wood substance is about 1.6 times as heavy as water, and this holds true of such light woods as poplar and basswood, as well as for heavy oaks or pine. Weight, therefore, depends upon the number of wood fibers and the thickness of their walls.

Weight is very important in identifying woods and in determining their usefulness for various purposes. Fuel value is in direct ratio with weight. The transportation of logs from the woods to the mill by driving is made possible by their relative floatability. This explains why driving cannot be practiced in logging hardwoods. In transporting logs or lumber, weight is an important factor in the cost of the operation.

The weight of wood is usually expressed in terms of specific gravity, that is, its relation to the weight of an equal volume of water weighed at a temperature of 4 degrees Centigrade. A cubic foot of pure water usually should be 62.43 pounds. Very few species of trees even approach this weight.

The following table gives the specific gravity of some of our common timbers according to Sargent:

Hardwoods.		Conifers.	
Species.	Specific gravity.	Species.	Specific gravity.
Lignum vitæ	1.14	Western larch	0.74
Mockernut hickory	0.84	Longleaf pine	0.70
Persimmon	0.79	Eastern larch	0.62
White oak	0.75	Loblolly pine	0.54
Hard maple	0.69	Douglas fir	0.52
Beech	0.69	Western yellow pine	0.47
White ash	0.65	Southern cypress	0.45
White elm	0.65	Sitka spruce	0.43
Black gum	0.64	Eastern helmlock	0.42
Black walnut	0.61	Amabilis fir	0.42
Red gum	0.59	Redwood	0.42
Black cherry	0.58	Lodgepole pine	0.41
Basswood	0.45	Incense cedar	0.40
Chestnut	0.45	Western white pine	0.39
Catalpa	0.45	Eastern white pine	0.38
Black willow	0.45	Balsam fir	0.38
Yellow poplar	0.42	Western red cedar	0.38
Butternut	0.41	Northern white cedar	0.32
Black cottonwood	0.41	Big tree (redwood)	0.29

The above list shows how much heavier, as a rule, the hardwoods are than the conifers, together with the great variation in weight in each class.

Shrinkage. — Moisture in wood is found in three distinct places; namely, in the cell contents, in the cell walls themselves, and in the intercellular spaces. Therefore when this moisture is released there is a consequent shrinkage in volume as well as in weight.

If wood was a uniformly homogeneous material, this shrinkage would be of minor importance because it would simply result in smaller size. But wood is made up of many complex elements. For example, there is heart- and sapwood, spring and summer wood, and pith or medullary rays, all of which vary in their tendencies in the process of drying.

When the top of a freshly cut board is exposed to the sun, it dries more rapidly than the lower surface with the result that there is greater shrinkage on the top surface and warping for a given area ensues. Another illustration is found in the fact that wood dries from the end more rapidly than from a side face; therefore there is shrinkage and frequently splitting or checking near the end due to the unequal drying process. Uneven or too rapid seasoning, therefore, results in checking, warping and twisting.

Wood shrinks very little along the grain and considerably across the grain. Again, a plain sawed board will shrink about twice as much in width as a quarter sawed board.

Roth has determined that for every hundred inches in width, boards of light conifers, such as white pine, spruce and cedar, will shrink about three inches; those of ash, walnut, beech, elm and maple about five inches; those of basswood, birch and chestnut about six inches; and hickory, young oak and eucalyptus will shrink from six to ten inches.

Cross-grained woods exhibit great irregularities in shrinkage and therefore considerable difficulty is experienced in seasoning eucalyptus, elm, black and red gum, beech and a few others.

Mechanical Properties.

Strength. — As applied to timbers, strength is a general term used in reference to the ability of wood to resist certain stresses. On investigation it is found that different woods vary quite materially in these respects.

The principal resistant features that go to make up strength are:

1. Resistance to compression along the fibers, as in the case of pillars, and dimension timbers in an upright position.
2. Stiffness or the ability to resist bending, as in the case of floor joists and beams supporting heavy loads.
3. Strength in tension or the ability to resist a lengthwise stress. Wood is seldom put to this test.
4. Shearing strength or the ability of the fibers to resist rupture either with or across the grain. As an example, the shear of a wooden pin in a mortise.

Moisture is an important factor in the strength of wood. Therefore, to a certain extent, strength increases with the degree of seasoning. Freshly cut or green timber consequently must be seasoned before being used wherever any strength factors are necessary.

Knots or other defects also influence the strength to a considerable degree. The size, character and location, however, of the knots are of importance. For example, in cross bending strength, knots on the upper surface of a beam do not detract from the strength nearly as much as on the lower part of the beam.

Weight of the wood is also important. Heavy woods are usually of strong structure.

The following are examples of commonly used woods of relative strength as expressed in its several different applications. The strongest all around woods are longleaf pine, larch, hickory, hard maple, yellow birch, white oak and black

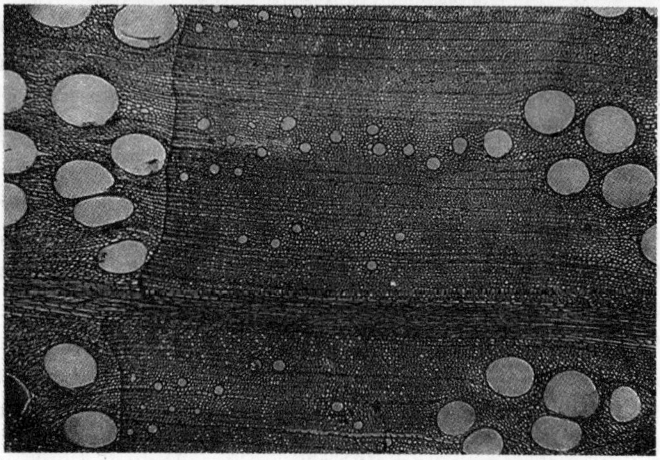

FIG. 42. — CROSS SECTION OF BLACK OAK (Quercus Velutina).

Magnified 50 diameters. This shows the large pores in the spring wood and the heavier and denser summer wood. Note parts of two growth rings and the line demarking them.

locust. Those of medium strength are cypress, Douglas fir, ash, beech, red oak, chestnut and sycamore. Some of our weakest woods, in this respect, are white and other soft pines, hemlock, spruce, basswood, yellow poplar, cottonwood and the western firs.

Hardness. — Woods vary a great deal in their hardness, which may be expressed as resistance to indentation or to the saw or axe across grain. Hardness is dependent largely on weight, structure of the wood elements and degree of seasoning. This feature is important in several lines of utilization, such as flooring, furniture, handles and many small wooden articles.

The following list shows the relative hardness of some of our more common species in the seasoned form.

Very hard.	Hard.	Medium.	Soft.	Very soft.
Hickory	Oak	Douglas fir	Western pine	White pine
Hard maple	Beech	Red gum	Hemlock	Sugar pine
Black locust	Birch	Tamarack	Chestnut	Spruce
Rock elm	Black gum	White elm	Yellow poplar	Redwood
Persimmon	Longleaf pine	Cottonwood	Cypress	Basswood
Osage orange	Ash	....	Cedar	Willow

Cleavability. — The resistance of wood to cleavage along the grain is important when it is desirable to split timbers. The line of least resistance in cleavage is along the radius because the medullary rays are in that direction. Wood splits much easier when wet because moisture softens the fibers and reduces adhesion across the grain. Straightness in the grain, however, determines to a large degree the ease with which wood splits.

The following list shows the relative splitting qualities of some of our woods:

Difficult to split.	Medium.	Easy to split.
Elm	Birch	Chestnut
Black gum	Maple	All pines
Beech	Hickory	Redwood
Sycamore	Oak	Cedars
Dogwood	Ash	Fir
Red gum	Cottonwood	Western larch

Altogether most of our woods are comparatively easy to split.

Miscellaneous. — Other mechanical properties that may be mentioned are flexibility and toughness. For example, hickory and ash are flexible, whereas hemlock and pine are brittle. Moisture content influences flexibility to a considerable degree.

Toughness refers to the combined strength and pliability of a wood. Good examples of tough woods are elm and hickory which offer high resistance to both tension and shearing.

Chemical Properties.

General. — As expressed before, wood is very complex in its structure and each species varies both in its physical and chemical characteristics.

The field of chemical utilization of our forest products has scarcely been touched, in so far as the possibilities are concerned.

Wood dried at 300 degrees Fahrenheit is made up of about 99 per cent of organic matter and about 1 per cent of inorganic matter. The inorganic elements make up the ash when wood is burned. At the above temperature, according to Roth, wood is made up of the following:

	Per cent.
Carbon	49
Hydrogen	6
Oxygen	44
	99

Among the other elements found to a small extent are nitrogen, potassium, sodium, calcium and magnesium.

Ordinary seasoned wood or lumber contains, in weight, about 25 per cent of water, 74 per cent of wood substance and 1 per cent of ash.

The wood itself consists of a skeleton of cellulose which contains in the different species varying degrees of lignin, tannin, resins, gums, etc. Cellulose and lignin are readily converted into starch and also sugar. Although an expensive method, sawdust is now being converted into sugar for animal food and it is already an important source of vinegar.

In the chapter on Utilization, the wood distillation industry is briefly described.

Durability. — As applied to wood, durability means the ability to resist decay or simply the length of life of a certain timber under given conditions. Durability is important in connection with those lines of usage where wood is especially susceptible to decay, such as railroad ties, poles, posts, mine timbers and piling. Durability may also include the influence of mechanical wear but this is relatively unimportant except in the case of ties.

The durability of the different species is very often the determinant factor in their value for several lines of utilization, especially in the case of timbers in contact with the soil, weather and water when used untreated.

Contrary to popular opinion, wood does not naturally decay. All decay in wood is caused by the work of fungi and bacteria which live on the starch and other material in the wood cells and cause the wood structure to break down, leaving the common dry rot, punk, blueing and rottenness in our timbers. The fungi are spread about by minute seed-like bodies, called spores, which are scattered about readily by the wind.

Fungi live and propagate and therefore decay is possible only whenever the following necessary conditions are present:

1. Sufficient heat.
2. Moisture in proper amount.
3. Oxygen.
4. Spores of proper fungi or bacteria.

Whenever any or all of these are removed, decay is not possible and, therefore, wood will last indefinitely. For example, some old piling that Caesar used in crossing certain rivers of France about two thousand years ago has recently been exposed and found to be in splendid condition. Irish bog oak is still sound because of the presence of acids in the logs and because air has been absent as it lay for centuries underneath the soil. Furniture seldom decays because moisture is absent and the wood finish prevents the entrance of fungi. For the same reason, we paint our houses and artificially inject poisonous antiseptic fluids into the wood fibers of ties, posts, poles, etc., to prevent the destructive work of the wood decaying fungi. A post rots at the surface of the ground first because at that point there are the greatest changes of moisture and heat. Sapwood is more susceptible to decay than heartwood because of its greater percentage of moisture and food for fungi and bacteria.

In the living tree the bark acts as a protective covering against decay, but whenever a fissure or crack is exposed or a limb is broken off, a splendid opportunity is offered for the entrance of the fungi.

There does not apparently seem to be any direct relation between the physical and mechanical properties of wood and its durability. As an illustration, weight, strength, stiffness, hardness or toughness do not seem to have any influence on the durability of any of our woods. Some of these properties, however, aid in the prevention of injurious effects of abrasion or mechanical wear. Two of our heaviest woods, hickory and hard maple, are not durable, whereas some of our lightest woods in weight, such as redwood, white cedar and catalpa, are very durable. Red cedar, a soft wood, is highly durable, whereas beech, a hard wood, is distinctly perishable.

There is, however, usually a definite relation between the

color of the heartwood and the durability of many of our woods. The darker the heart, the more durable is the wood, especially in the case of ebony, lignum vitæ, catalpa, red cedar, black locust, osage orange and several others. Many of our species with light colored heartwood, such as basswood, maple, hickory, spruce and tupelo, are very perishable.

The durability of any species depends, therefore, on certain chemical constituents, such as resins, gums, tannin and other decay resisting materials. These happen to give a dark discoloration to the heartwood of several of our species so that this explains the relation of color to durability.

Rapidity of growth, within a species, is an important factor in its durability, but as between species, there is no direct inference. As an example, rapid growing species, such as black locust, chestnut and catalpa, are durable as well as such slow growing species as cypress, longleaf pine, western larch and the cedars.

The following table shows the relative durability of some of our more common species:

Very durable.	Durable.	Intermediate.	Perishable.	Very perishable.
Black locust	White oak	White pine	White elm	Black gum
Red cedar	Black ash	Norway pine	Beech	Basswood
Live oak	Cherry	Shortleaf pine	Hickory	Buckeye
Black walnut	Red elm	Red oak	Hard maple	Paper birch
Cypress	Persimmon	Red ash	Red gum	Aspen
Western red cedar	Longleaf pine	Yellow poplar	White ash	Willow
Redwood	Western larch	Butternut	Loblolly pine	Sycamore
White cedar	Eastern larch	Sugar pine	Hemlock	Lodgepole pine
Lawson cypress	Cuban pine	....	Spruce	Balsam fir
....	Ironwood	....	Yellow birch	Jack pine

BIBLIOGRAPHY.

BROWN, NELSON C. An Analysis of Durability. Hardwood Record, March 25, 1913.

CLINE, MCGARVEY. Strength Values for Structural Timbers. U. S. Forest Service Circular 189.

FERNOW, B. E. Timber Physics. Parts I and II, U. S. Forest Service, Bulletin 8.

FOREST SERVICE CIRCULAR 15, Summary of Mechanical Tests on Thirty-two American Woods.

RECORD, S. J. Economic Woods of the United States. John Wiley and Sons.

RECORD, S. J. Grain and Texture in Wood. Forestry Quarterly No. 1, Vol. IX, 1911.

RECORD, S. J. The Mechanical Properties of Wood. John Wiley and Sons, 1914.

ROTH, FILIBERT and B. E. FERNOW. Timber: An Elementary Discussion, etc. Division of Forestry, Bulletin 10.

CHAPTER XII.

WOOD PRESERVATION.

History and Importance.

Ever since ancient Roman times, various methods of artificially treating wood to prolong its life have been used. At first charring timbers to prevent decay in the ground was used. Painting the surface of the wood was also frequently resorted to.

The idea of injecting chemical preservatives into the wood fibres dates back to 1657 when Glauber, a German chemist of Carlstadt, carried on the first successful experiments. Since that time, the industry has taken wonderful strides and many methods are now in use to prolong the life of timbers most subject to decay, such as ties, posts, poles, piling, mine timbers and construction timbers, wood blocks for paving, etc.

In this country, the first successful timber treating plant was established at Lowell, Mass., in 1848, and it is still in operation.

Wood preservation is really a means of more intensely utilizing the products of the forest, because if we can increase the life of certain timbers by artificial and reasonably cheap methods, it will decrease the demands for wood supplies and therefore play an important part in our forest economy.

The annual loss of wood supplies due to decay amounts to over seven billion board feet or over 70 per cent of all the causes of wood destruction. Another pertinent fact is that our more durable species, such as white oak, cedar, black locust, and cypress, are rapidly disappearing and we are being forced to use inferior species which, in many cases, by a cheap preservative treatment, can be made to last longer than the more durable species in their natural condition.

In 1885 there were only three pressure plants in this country.

The importance of the industry can be at once seen from the following amounts, conservatively estimated, that are now being treated annually:

Class.	Amount.
Ties	33,000,000 pieces
Paving blocks	2,800,000 square yards
Poles	2,000,000 lineal feet
Piling	11,000,000 lineal feet

Besides this, large quantities of fence posts, mine timbers, cross arms and construction timbers are treated every year. Ninety per cent, however, of treated timbers is composed of railroad ties, expressed in volume of wood.

This industry obviously, then, consumes enormous amounts of chemical preservatives. The favorite all around chemical is creosote oil, a product of the distillation of either coal or wood tars. In 1912, 83,000,000 gallons of creosote were used in this industry besides about 20,000,000 gallons of zinc chloride and 3,000,000 gallons of other preservatives among which may be mentioned crude petroleum, copper sulphate, mercuric chloride, several patent chemicals and others of less importance.

The measure of good preservative may be summed up as follows:

1. It must be available and reasonably cheap.

2. It must be antiseptic and poisonous to wood decaying fungi.

3. It must not readily evaporate or leach out of the wood fibers.

Principal Methods.

The methods pursued in wood preservation may be classified as follows:

1. Pressure or cylinder processes.
2. Open tank process.
3. Brush treatment.

The impregnation of wood by injecting chemical preservatives into the wood under pressure in large cylinders is the method used with ties, paving blocks and to some extent with other forms of timbers. It is, therefore, by far the most important. The open tank treatment is used with poles and posts usually for treating that portion most subject and liable to decay. The brush treatment is largely applied to shingles, posts, poles and farm timbers and is relatively of little importance.

Pressure or Cylinder Process. — There are a great many variations of the pressure process, depending largely on the preservative used, its adaptability to the species treated, the penetration desired, etc. Many of them have been patented and are known by trade names.

The principal and most common method followed, however, is the Bethell or Burnettizing process. When used with creosote, it is called the Bethell process, and with zinc chloride, it is called the Burnettizing process. The following is a brief description of the method.

The timber to be treated, usually ties, is first seasoned for from three to ten months. It is placed on iron trucks, called "cylinder buggies," and pushed on the tracks from the yards directly into huge horizontal cylinders about six to eight feet in diameter and up to one hundred and fifty feet in length. These cylinders are constructed to withstand high pressure and the heavy doors are hermetically sealed. At first live steam is introduced into the cylinder and a pressure of about

FIG. 43. — HIGH-PRESSURE CYLINDERS USED IN TREATING TIES, OAKLAND, CALIFORNIA.

Ties are run into the cylinders on trucks and impregnated with creosote or other preservatives to increase their durability.

twenty pounds per square inch maintained for several hours, depending on the species, their size and moisture content. This steaming softens the wood fibers and opens up the pores in the wood. The steam is then led off and a vacuum applied. This exhausts the air in both the cylinder and the wood structure itself and prepares the way for the penetration of the preservative fluid. The vacuum is maintained for about one-half hour, after which, without reducing the vacuum pressure, the preservative, either creosote or zinc chloride, is run into the cylinder at a temperature of from 170 to 190 degrees Fahrenheit and pressure pumps exerted to force the fluid into the wood fibers. The amount of fluid injected depends upon the species and the amount desired in the tie. The pressure is then released and, in some cases, a vacuum is again applied to draw out the excess fluid and hasten the drying process. The ties are then drawn out of the cylinder on the trucks and piled in the yards to dry out until wanted in the tracks.

The Rueping, Card, Lowry and other more or less important processes are in common use but they are all variations of the same pressure treatment.

The following table published by the Forest Service illustrates to best advantage the results of the treatment of railway ties as compared with untreated service.

ESTIMATED LIFE OF UNTREATED AND TREATED TIES.

Species.	Untreated life in years.	Treated with 10 pounds of creosote per cubic foot, years.	Treated with 0.5 pound of zinc chloride per cubic foot, years.
Longleaf pine	7	20	(not used)
Chestnut	7	14	11
Douglas fir	6	15	11
Spruce	6	14	11
Western pine	5	17	12
White pine	5	14	10
Lodgepole pine	5	16	11
Tamarack and hemlock	5	15	11
Red oaks and beech	4	20	12
Maple	4	18	12
Gum	3	16	11
Loblolly pine	3	15	10

Such naturally durable woods as black locust, redwood, cedar, cypress and white oaks are not shown in the above table because they will last under average conditions for from eight to twenty years and are therefore seldom treated. All the above estimates are based on the use of tie plates in the maintenance of the railway track.

The cost of treating ties varies with the method used, the species, character of heart and sapwood and degree of penetration. With some species both heart and sapwood are easily saturated with a full penetration; with others, only the sap can be successfully treated.

The cost of treating with creosote is usually about twice that with zinc chloride but the extra cost is evidently justified in the extra service rendered by creosoted ties in the long run. For instance with maple, using the above tabulated figures, the cost of creosote treatment is 37 cents per tie and of zinc chloride only 17 cents, but the annual saving over untreated ties is 0.136 cent per tie with creosote as against 0.126 cent for zinc chloride, taking into consideration the original cost, as well as the cost of treating the ties. Zinc chloride, however, is highly successful in dry regions where this preservative does not leach out.

Open Tank Process. — The open tank treatment is used mostly with posts and poles. Farmers and users of electric light and telephone poles are finding that the prices of the more durable species are advancing so rapidly that they are turning to the use of perishable or inferior species and getting very satisfactory returns by a simple and cheap preservative treatment.

The open tank process is used almost entirely with creosote and the following method is usually pursued. The bark is first removed and the wood thoroughly seasoned. This seasoning process replaces to a considerable extent the moisture in the wood cells with air. The posts or poles or that portion to be treated (usually a distance slightly in excess of

FIG. 44. — A POST TREATING PLANT, MONTGOMERY CO., MARYLAND.

500 million new fence posts are required on our farms every year. Creosote treatment costing from 6 to 10 cents per post will often double the life of posts in the ground.

its depth in the ground) is immersed in a hot bath of creosote up to a temperature not exceeding 215 degrees Fahrenheit. This heating process lasts for from two to six hours, depending on the species and their size. During the heating the moisture and air in the wood expand and a good portion of them pass out, appearing as steam or little air bubbles at the surface. The posts or poles are then hurriedly removed to a cold bath of creosote and the contraction of the air and moisture in the wood, due to the cold, creates a partial vacuum, which is destroyed by the entrance of the preservative fluid.

In this way atmospheric pressure, due to change in temperature, accomplishes the impregnation which is secured to a much greater degree in the large cylinders by artificial pressure.

Inasmuch as sapwood is more readily subject to treatment than heartwood, due to its peculiar structural and chemical properties, species with a large percentage of sap and timbers in the round are treated much more easily than when split.

The open tank treatment is usually carried out in a very simple and cheap way. It is especially adapted for use by farmers or a group of them who can use one centrally located plant. It usually consists of a galvanized iron tank or container over a bricked-up oven. A fire is built in the oven for heating purposes and a smokestack is erected to carry off the smoke and create a draft. About 500,000,000 new fence posts and 3,500,000 poles are used every year in this country so that it is becoming an important matter to prolong the life of these timbers.

The cost of treatment per post is from 4 to 12 cents apiece, including cost of equipment. The cost of treating poles is about $1.50 apiece for the average pole. The results from this treatment are certainly more than justified. The average increased life of the treated post is fourteen years and of the pole about ten years.

Many of the most durable species are not treated, especially when used for posts. Among these may be mentioned osage,

FIG. 45. — A SMALL BOUCHERIE TREATING PLANT IN A CALIFORNIA NATIONAL FOREST.

The Boucherie process is used to impregnate telephone and telegraph poles.

orange, red cedar, black locust, mulberry, catalpa and occasionally white oak and white cedar.

The following tables, partly taken from the Forest Service, illustrate the estimated increased life, in years, to be expected from treated posts and poles of a few typical species:

POSTS.

Species.	Untreated life in years.	Treated life in years.	Increased life as result of treatment.
Longleaf pine	8	20	12
Douglas fir	7	20	13
Red oak	5	20	15
Tamarack	5	20	15
Chestnut	10	20	10

But perhaps the greatest economy can be effected with the use of such perishable and cheap post material as cottonwood, elm, maple, birch, beech, willow and poplar which can be made to last from fifteen to eighteen years instead of from three to five years.

POLES.

Species.	Untreated life in years.	Treated life in years.	Increased life as result of treatment.
Chestnut	8–10	20	10–12
Lodgepole pine	5	20	15
Western pine	6	25	19
White oak	8	20	12
Douglas fir	8	20	12

Brush Treatment. — The brush treatment is a very cheap, but less efficient method of prolonging the life of certain timbers, especially those exposed to the weather or even in the case of poles and posts when the open tank treatment cannot be used.

There are several patent preservatives on the market which are very good. Any of these, or creosote applied hot and with at least two coats, often give fairly satisfactory results. The ordinary paint brush is used to apply the preservative. Although only a superficial treatment, the object should be to thoroughly fill and cover with the fluid all cracks, checks, knot holes or similar defects which offer the best opening for the wood-destroying fungi to work in. All wood should be thoroughly air-dried before applying the liquid. Dipping is really a form of the brush treatment and gives excellent results with shingles, posts, sills and other timbers.

Results of Wood Preservative Treatment.

The following table presents a summary of the results that may be expected from wood preservation as applied to the principal timbers by the most effective method in each case.

Class.	Life untreated, years.	Life treated, years.	Average cost of treatment.	Increased life in years.
Ties	7	17	$0.35	10
Poles	13	23¼	1.50	10¼
Posts	8	22	0.10	14
Piles	3½	21½	0.25 (per cu. ft.)	18
Mine props	3	13	0.11	10
Lumber	18	32	10.00 (per 1000 board ft.)	14
Shingles	8	20	.....	12

BIBLIOGRAPHY.

KEMPFER, W. H. The Preservative Treatment of Poles. U. S. Forest Service Bulletin 84.

PETERS, E. W. The Preservation of Mine Timbers. U. S. Forest Service Bulletin 107.

SHERFESEE, W. F. A Primer of Wood Preservation. U. S. Forest Service Circular 139.

SHERFESEE, W. F. Wood Preservation in the United States. U. S. Forest Service Bulletin 78.

SMITH, C. S. Preservation of Piling against Marine Borers. U. S. Forest Service Circular 128.

WEISS, H. F. The Preservative Treatment of Fence Posts. U. S. Forest Service Circular 117.

WEISS, H. F. Prolonging the Life of Cross Ties. U. S. Forest Service Bulletin 118.

WILLIS, C. P. The Preservative Treatment of Farm Timbers. U. S. Dept. of Agric. Farmer's Bulletin 387.

CHAPTER XIII.

FOREST ECONOMICS.

Owing largely to the superb forests of virgin timber awaiting the axe of the first settlers, the civilization of North America has been founded, to a remarkable degree, on wood. The primeval forest contained, approximately, 850,000,000 acres, and to the marvelous richness of the original forest is due the prodigal extravagance of the American nation. As stated in Chapter I, the annual per capita consumption of timber in the United States, 250 cubic feet, is more than six times that of Germany and more than twenty times that of Great Britain.

Economic Value of Forests.

Next to food in human economy, shelter is most important and, according to Fernow, over half our population live in wooden houses and two-thirds of the population use wood for fuel. The same author estimates that fully 95 per cent of all the timber consumed is for necessities.

In spite of the advances in the structural arts, timber is increasing in demand, the per capita consumption being twice what it was fifty years ago. Steel, concrete and brick may be substituted in some cases, but even the manufacture of these substitutes requires considerable wood and, in addition, only 25 per cent of the lumber output may be considered as structural timber. The attempt to use concrete ties in Germany, where trains must be run at high speed, has met with poor success. Where heavy traffic at slow speed is the rule, concrete ties may serve, but the rigidity and lack of elasticity detract greatly from the comfort of the traveler and from the durability of the rolling stock. Viewed from every side, it is extremely difficult to find any material natural or artificial which can take the place of wood.

Minor products, like shoe pegs, spools, musical instruments, etc., consume an astonishing amount of timber. New York State alone consumes over 8,000,000 board feet per annum in the manufacture of wooden novelties, and over 6,000,000 board feet are required annually by the manufacturers of clocks and toys. Houses, furniture, ship yards, railroads, newspapers and books all require wood. A shortage in the timber supply would be felt almost as keenly as a famine in foodstuffs.

Aside from furnishing wood, forests may serve other purposes. In certain parts of the West, the function of the forest in controlling run-off, thus gradually releasing the water for irrigation purposes, may be paramount to furnishing local timber. The retarding of surface waters to prevent erosion, the checking of drying winds in the prairie regions, the influence of forests on climate, are all indirect influences of great importance. Indeed, a country without at least twenty per cent of its land area covered with forest, is at a decided disadvantage from an economic and climatic standpoint.

Forest versus Agricultural Soils.

As explained in previous chapters, Forestry is concerned primarily with lands unsuited to agriculture, but nevertheless, capable and sincere men may disagree on what constitutes true forest soil. In eastern Texas, for example, efforts are being made to colonize land from which the longleaf pine has just been cleared. Agriculture or horticulture are doomed to failure on such sites, because of the sterility and excessive drainage, yet the promoters are absolutely sincere. By experience alone can the exact worth of the land be proven, and the proper adjustment of soils to transportation facilities and economic conditions will only be reached after years of trial.

Future Use of Land. — The total area of the United States is 3,026,789 square miles or, approximately, 1,900,000,000 acres. Of this, about 550,000,000 acres are now covered with forest (65 per cent of the original forest area*), and 415,000,000 acres are devoted to agriculture.

Fig. 46. Forest Fire Burning in the Debris of a Previous Fire.

Unless the fire-killed timber can be sold and removed within a short time, it is not only a total loss financially, but may feed a second fire of even greater severity.

Within the next fifty years, the land, suitable for growing forests, will decrease to about 360,000,000 acres, or less than twenty per cent of the area of the United States. About 90,000,000 acres will be held as farm woodlots, bringing the total area devoted to timber production up to 450,000,000 acres.

Of this acreage, 63 per cent will be in the West and 37 per cent in the East, and from this diminished area, it will be necessary to supply the timber needs of a population, conser-

* See Appendix for table.

vatively estimated at 150,000,000. Any hope of obtaining timber, from any of the nations now exporting timber, is futile, as the surplus of such countries as Russia, Finland, and Sweden will be quickly absorbed by continental countries to supply the local demand.

With a population rapidly gaining and a fixed area, the only solution is to put each acre of land to its best permanent use. The Federal Government has made a start with the National Forests and this example should be extensively followed if the future American citizen is to enjoy many of life's necessities.

Ownership of Forest Land. — Before a rational forest policy can be started under a free government, the owners must be convinced that management in perpetuity is financially profitable. In certain of the German states, the cutting of timber on private lands is regulated by the central government. In spite of advanced legislation passed by certain states, it is doubtful if state regulation will ever be enforced to that degree in the United States. The rights of the individual are too strongly entrenched; it is contrary to the principles of democracy. The problem then resolves itself into educating or persuading the individual owner to handle the forest land for present profit without endangering the future timber supply.

According to the U.S. Bureau of Corporations, the present status of timber ownership in the United States is as follows:

Private	75.0 per cent
National Forests	21.5 per cent
Other, Federal and State	3.5 per cent
	100.0 per cent

The same bureau reports some interesting discoveries concerning the ownership of land in general; forest land constitutes a large portion of the corporate holdings investigated. To quote from its findings:

"The control of our standing timber in a comparatively few enormous holdings, speculatively held far in advance of any use thereof, and the great increase in value of timber, resulting in part from such speculative holding, are underlying facts that will become more and more important elements in determining the price of lumber as the supply of timber diminishes," says the report.

"The main fact shown is that 1694 timber owners hold in fee over one-twentieth of the land area of the entire United States from the Canadian to the Mexican border. In many states these 1694 own no lands at all. In the 900 timbered counties investigated they own one-seventh of the area.

"These 1694 holders own 105,600,000 acres. This is an area four-fifths the size of France, or greater than the entire state of California, or more than two and one-half times the land area of the New England states. Sixteen holders own 47,800,000 acres, or nearly ten times the land area of New Jersey. Three land grant railroads own enough to give fifteen acres to every male of voting age in the nine western states where almost all their holdings lie."

Further comment is made to the effect that such marked concentration has a decided effect on the future supply of timber, ore, gas, water powers, etc., and that future distribution of these holdings is by no means probable. Out of 7,400,000 acres originally granted to railroad, wagon road and canal corporations, only 15 per cent are now distributed in small holdings. In 1910, three western railroads still held 40 per cent of the 82,000,000 acres granted them in 1865–1870. Such control of land and forest resources makes monopoly easy of accomplishment, and the Federal ownership of one-fifth the standing timber is the best protection against future exorbitant prices. As described in Chapter I, the Forest Service is annually selling over $1,500,000 of stumpage, but this amount represents a fraction of the total stand; it is merely the mature timber that is accessible which is now marketed, and with the development of better transportation facilities, the more remote portions of the forest will become available and the supply of timber owned by all the people of the United States will be a potent factor in controlling excessive prices.

Forest Resources of the United States.

The forests of the United States have been estimated by the Forest Service to contain 2,500,000,000,000 board feet, a reduction of 2,700,000 000,000 from the original forest. Of this, 75 per cent is privately owned, the distribution of which is as follows:

Pacific Northwest	46.0 per cent
Southern Pine region	29.1 per cent
Lake States	4.5 per cent
Other regions	20.4 per cent
	100.0 per cent

By species, the following estimate of the standing timber has been compiled by the Forest Service:

Species.	Billion board feet.
Douglas fir	525
Southern yellow pine	350
Western yellow pine	275
Redwood	100
Western cedar	100
Western hemlock	100
Lodgepole pine	90
White and Norway pine	75
Eastern hemlock	75
Western spruce	60
Eastern spruce	50
Western firs	50
Sugar pine	30
Cypress	20
Other conifers	100
Hardwoods	500
Total	2500

This stand of timber, enormous as it may seem, can by no means be considered inexhaustible, for since colonial times, it is estimated that timber exceeding the present stand by two billion board feet has been used, wasted or consumed by forest fires. In view of the low growth rate and the excessive annual drain, the present forest capital is none too large.

Rate of Consumption. — In the chapter on Utilization, the annual production of lumber was stated to be 40,000,000,000 board feet. This amount, together with the timber used for ties, poles, posts, firewood, etc., makes the total annual consumption of forest products amount to 20,000,000,000 cubic feet of wood. On the basis of saw timber alone, the present supply should last about sixty years without taking growth into consideration. (The average annual increment per acre of forest land, as a whole, in the United States is about one-fourth that obtained in German forests and, in addition, 200,000,000 acres are estimated to be in over-mature or virgin forests where growth is offset by decay.) Unfortunately, however, there is not an equal demand for all kinds of timber, and there is no doubt but that certain species will command very high prices within the next thirty or forty years. There will be no real timber famine, however, in the ordinary sense; timber of some kind will be available, but the quality may be much inferior to the average grade now used and the prices will range considerably higher. In order to prevent any hardships resulting from a diminishing timber supply, some readjustments will be necessary. The excessive per capita consumption of forest products every year (250 cubic feet) must be greatly reduced; closer utilization at the mills and factories must be practiced, so that more than 40 per cent of the tree is actually used and, finally, better forestry methods must be put in force to triple or quadruple the present per acre yield of the non-agricultural land. Before the demand can be levelled down to the reduced supply, the pinch of timber poverty may be keenly felt.

Methods of Providing Future Supply.

In view of the ownership of forest resources in the United States, the several agencies must co-operate to provide the coming generations with forest products. The Federal and State Governments own only a fourth of the present stand, but by husbanding their resources and regulating its use, the publicly owned timber supply may exercise an influence far greater than its amount would indicate.

State versus Federal Control. — To secure ideal management of forest land, certain conditions are quite essential.

1. There should be a large area managed as a producing unit. This will permit better regulation of yield, closer utilization (specialization) and reduction of overhead charges.
2. The plan of management should be made with regard to future conditions and should not be subjected to radical changes at frequent intervals.

The above conditions can best be met by having the land owned by an organization with perpetual existence like the state, government, or a corporation, for forestry at best is a long-time investment. No private corporation is likely to forego present profits in order to reap future returns, so the future welfare of the country can best be served by having as much timber land as possible owned by the government. Either the State or National government can dispense with present revenues to assure the future generations an adequate timber supply and, in the final analysis, a good government is responsible for the welfare of its citizens, both present and future. Which method of controlling natural resources, State or Federal, will best serve to promote their best use, is still a mooted question and one which has provoked much discussion.

Since the first National Forests were withdrawn from the public domain, there has been a continuous attack on the Federal forest policy in certain parts of the West. These attacks may generally be traced to one of two sources; either to corporations interested in mining, lumbering, or water-power development, or to zealous partisans of the doctrine of states' rights, who believe the state should control all the resources within its boundaries, and refuse to admit the possibility of better control or regulation by the central government. Such well meaning citizens are often used to advantage by corporate interests, and at the last Conservation Congress in Washington, men, whose sincerity and integrity were above suspicion, supported the demand of individuals with ulterior motives, that the present National policy should be reversed, and that all natural resources should be turned over to the separate states.

The fact that public ownership of forest land is extremely important to the welfare of future citizens has been previously brought out; but why should not State ownership prove successful? Federal ownership has the following advantages:

1. While congressional action is slower and harder to secure, when once favorable laws are enacted, their repeal is much more difficult than with state legislation.

2. A Federal organization, while slightly hampered by its size, can, nevertheless, develop a corps of specialists impossible in a state service.

3. Interstate questions can be settled to advantage by the central government.

The states should be encouraged to acquire land for state forests, but, on the whole, Federal regulation of all natural

FIG. 47. AGRICULTURAL LAND RUINED BY FLOOD—MARION, McDOWELL CO. NORTH CAROLINA.

Aside from their importance in producing timber, the value of forests in retarding run-off is very large. This land formerly worth $125 per acre is now useless owing to gullies and heavy deposit of gravel caused by flood.

resources including forests, is best calculated to achieve the ends of a government for the people.

The arguments that are most often used by the opponents of the present forest policy are that the state is abundantly able to look after its own welfare, and, since the eastern states were not hampered in the disposition of their natural resources, the same privilege should be accorded the younger states.

The ease with which practically any state legislature reverses itself, is a sufficient argument against the first contention. Many instances are available where states have bartered their rights and resources for a song. Of all the states given land for school purposes, very few have handled these possessions with much foresight. In one of the western states, where opposition to the Forest Service has been especially keen, the legislature sold state land containing billions of board feet of virgin white pinc to a syndicate at $1\frac{1}{4}$ cents per thousand board feet. The Federal government is now selling the identical timber in the same region for a price ranging from $4 to $5 per thousand board feet.

Concerning the propriety of demanding equal privileges with the older states, little need be said. Because one region has squandered its patrimony, is no reason why the same license should be granted to another, especially when much of the Public Domain was purchased from funds supplied by the original states.

Under present conditions, Government ownership is no bar or hindrance to legitimate development; agricultural land can be acquired in compliance with the Act of June 11, 1906, mining claims can be entered, timber bought and water powers developed. The officials in charge insist, however, that the rules be observed, and that the property of the nation be used without impairment. Free and proper use is encouraged, but the monopolistic acquisition or control of resources for speculation is opposed. This attitude has aroused strong hostility in certain parts of the West.

State Forest Work.* — At present, thirteen states have a commission or bureau concerned with forestry alone; fifteen additional states have a joint conservation, or forest and game commission and in three states, there is a single commissioner, making thirty-one states with a clear-cut forest policy and proper machinery for the enforcement of their statutes. Twenty of these states have one or more technically trained foresters in charge of their work.

The above states own 142 State forests, containing 3,426,832 acres; they operate 30 forest nurseries, comprising 150 acres with a yearly output of nearly ten million seedlings. Sixty-three forest experiment stations are operated by eleven states, and already over 12,000 acres of forest plantations have been started. In such states as Minnesota, Michigan, Maine, New York and Pennsylvania, there is abundant room for further development of a vigorous forestry policy. New York, for instance, owns 1,800,000 acres of wild land, lying in the Adirondacks and Catskills, containing 25 per cent of the standing timber of the state. With one-tenth of the population of the United States living within its boundaries, provision for future supplies of timber must be made. Other states are in a like situation, and where there are vast areas of unseated land, due to lumbering, fire, etc., the creation of state reserves from such lands serves to restore equilibrium to the land situation. Practically every state which has purchased wild land to any extent has found it a splendid investment on account of the increase in value and, in addition, the values of adjoining properties show a decided price appreciation. The primary object of establishing future forests and recreation grounds is gained and a situation which

* For tabular statement see Appendix.

has been extremely serious in some states is vastly improved.

Private Practice of Forestry. — The adoption of intensive private forest management on any scale has not gained much headway in the United States, although statistics compiled by the Forest Service indicate that over 3,500,000 acres can be considered as receiving some extensive form of treatment. The absence of law of entail has a tendency toward the partition of landed estates so that the bulk of large original grants is now held in small parcels. Certain exceptions to this may be found in different regions; the heirs of an early Maine settler now control and lumber 230,000 acres; several tracts of 60,000 to 100,000 acres are found in the Adirondacks and the Biltmore forest of 120,000 acres, situated near Asheville, North Carolina, are instances of successfully managed private forests. The latter estate is especially interesting since it has been in charge of technically trained foresters for the past twenty years, and its recent purchase by the Government will secure for all time the benefits of some very instructive experiments.

Forestry, as a rule, does not appeal to the individual on account of the time element in securing returns. In Great Britain and on the continent, where large estates have been in possession of the same family for generations, the forest land may furnish a considerable revenue. In the United States, individuals prefer an investment which is quicker to mature and is more negotiable than forest land. However, if forest capital in the form of mature timber can be purchased at a reasonable figure, conservative lumbering with an eye to future crops will prove highly profitable.

The type of land owner most advantageously situated concerning the practice of forestry, is the farmer owning woodland in connection with farms, because the tillable land yields an annual return and no forced cuttings are necessary. The woodlot furnishes domestic timber and fuel, thinnings can be made during the winter when other work is slack and, on the whole, a farm with a good woodlot is much more attractive and desirable as a home and as an investment. The practice of intensive forestry by the farmer is of prime importance, and the education of and co-operation with this type of owner should be heartily encouraged. The most discouraging feature of woodlot forestry in the past has been the extremely low returns received for stumpage, cordwood, etc. By assisting the small owner to market his forest products at a profit, treatment accorded the 190,000,000 acres composing the nation's woodlot will be much better in the future than in the past.

Communal Forestry. — In a nation as young as the United States, the ownership and management of any land for the benefit of the community, as a whole, is hardly to be expected, especially when forest resources have been so abundant. Solidarity can only be acquired after years of leisurely and harmonious community existence. Most of the American towns are still too busy growing to think of developing resources for the future citizens, consequently, examples of communal forestry are still quite unique.

European countries furnish countless examples of non-agricultural land owned and operated for the benefit of all the citizens. The returns from some of those forests are strikingly large. The Sihlwald, the City Forest of Zurich, Switzerland, has been owned by the city for upward of 1000 years. In spite of spending about $4.50 per acre per year in tending the forest, the annual net revenue per acre amounts to approximately $7.50.

Gaulsheim, a small village in the Black Forest, pays the operating expenses of the village out of the revenues obtained from the forest of 2000 acres. In favorable years, it has paid a dividend of $4 to each of the 800 inhabitants. Such high returns are hardly possible in view of the lower timber values and higher labor prices that obtain in the United States. If communities would realize the possibilities that non-agricultural land offer in the way of revenue production, many of the unattractive waste areas lying close to our towns and villages would be purchased and planted to the æsthetic and economic improvement of the community. Reforestation of city water sheds has great possibilities in this connection. Grazing, or tilling and fertilizing the soil are impossible on account of contaminating the water supply. A crop of trees, on the other hand, improves the water holding capacity of the soil; the roots, on entering the ground, form basins in which the water may collect during a heavy rain and, in many ways, the presence of forest cover on a water shed, is most desirable. It is the one profitable and hygienic use to which such land may be put.

Up to the present time, there are 97 municipal forests in 13 states, Massachusetts leading with 56. Such forests, for the most part, may be used for recreation grounds as much as for timber production. The start has been made, however, and the extension of municipal forestry will prove an important phase of the general forestry development of the next twenty or thirty years.

Æsthetic Forestry. — As defined in the first chapter, forestry, strictly speaking, is the raising of trees in forests for timber purposes, while arboriculture is concerned with the growing of trees singly or in groups for any purpose whatever. Thus, æsthetic forestry is a rather contradictory expression which, nevertheless, has gained more or less sanction through use.

Æsthetic forestry really divides itself into what might be called park forestry, and shade tree work. In the former line of work, forestry and landscape engineering are combined; in the latter, specialized silviculture is most important. In both, an intimate knowledge of the habits and resistance of trees, their form and growth rate, their peculiar qualities, rendering them suitable for one location and unfitted for another, is necessary.

The importance of so-called æsthetic or city forestry has been, to a large degree, overlooked by many technical foresters. In a country where public opinion has such an important bearing on the enactment and enforcement of laws, any line of activity, awakening interest in single trees, will have its effect in stimulating a greater interest in trees in general. Thus, the psychological importance of shade tree work and æsthetic forestry is extremely great. Interest a citizen in the shade tree before his house or on his lawn, and a long step has been taken toward awakening intelligent interest in and support of economic forestry. Favorable public opinion and interest are necessary to secure appropriations for economic forestry and, as an entering wedge, æsthetic forestry has enormous possibilities.

BIBLIOGRAPHY.

FERNOW, B. E. Economics of Forestry.

KELLOGG, R. S. The Timber Supply of the United States. U. S. Forest Service, Circular 166.

KELLOGG, R. S. Lumber and its Uses.

VAN HISE, CHARLES R. The Conservation of National Resources in the United States.

ZON, RAPHAEL. The Future Use of Land in the United States. U. S. Forest Service, Bulletin 159.

SUNDRY ARTICLES. Report of National Conservation Commission, Volume II, pages 178–196.

CHAPTER XIV.

FOREST FINANCE.

Definition and Discussion.

Aside from furnishing material which is indispensable to our modern civilization, controlling the run-off of precipitation, etc., economic forestry demands that forests shall also yield a cash revenue, for unless forestry proves attractive financially, its practice will be confined to the government or other agencies which can afford to supply timber at a loss because of indirect benefits. Forest finance deals with the question of determining the profitableness of growing timber and in the final analysis, therefore, is of importance both to the state and individual.

The principles involved in forest finance are the same as those employed by insurance companies and bankers, except that the foresters of the United States have not a good set of "yield tables" available. The reactions of capital and interest are easily computed, but without such tables, the future volume yield can only be approximated. However, the results obtained in other countries where forestry has long been established are of great assistance and give some conception of the outlook in the United States. The value of forest finance as a check in regulating the relative intensiveness of forestry operations can hardly be overestimated.

General Considerations.

Capital and Interest. — The needs of the nation create various demands and capital furnishes the means of satisfying them. While labor or a certain kind of material may satisfy a given demand, money is the usual standard of material values.

Interest is the earning power of capital, the rent paid for the use of the principle. The interest rate asked will largely depend on the demand for money, negotiability and safety of the investment.

The general laws affecting capital which have the most bearing on forest finance are as follows:

1. The value of any natural resource tends to rise as the population increases. Timber values are much higher now than twenty-five years ago; as a rule they may be said to have increased at the rate of twenty per cent a decade, during the past forty years.

2. The ready negotiability of capital adds to its value. At present forest land is not readily sold nor can money be borrowed upon it as easily as on government bonds for instance. But with the rapid depletion of the stands of virgin timber, not only will the cash value of timber vastly increase, but also the ease with which it may be sold or used as collateral.

Interest is of two kinds, simple and compound. Compound interest is used chiefly in forest finance, since the forest crop is not harvested annually, as in farming, but at the end of a long rotation. In addition, capital and interest are hard to separate since the growth-ring (interest) is annually laid on top of the capital and previous interest, represented by the tree.

The laws governing compound interest overlook any possible loss of the entire capital, since the total returns compounded assumes the safe delivery of the original investment and interest. Neither is there supposed to be any change in the interest rate during the rotation. Thus, since any accidents, like loss from fire or windfall, or decrease in the growth rate, owing to insect attacks, etc., are not considered in the computations, additional allowance must be made.

Forests as Investments. — Only within the last two or three decades has the idea of holding forests as investments gained any standing in the United States. Timber land for the most part had been regarded as a natural resource, incapable of growth or renewal; it contained a certain amount of raw material and like a gold mine, the quicker these assets could be converted into cash, the better for the owner or stock holders. Of late years, however, the fact that forests are increasing in value and that, if properly handled, they will regenerate themselves naturally and grow at a profitable rate, is becoming appreciated, and large tracts of immature forests are being bought up for speculation purposes.

Stumpage is bound to rise, not only because of the decrease in supply but the better transportation facilities, the greater safety, owing to improved fire protection, and the increased negotiability of woodlands, will greatly stimulate the demand for forest land and forest products.

Aside from the speculative value, forest land will prove decidedly attractive to certain groups of corporations and individuals. Entailed estates, water and power companies, paper concerns, towns and communities will all turn to forests as a conservative investment, yielding a fair return, but the rate will not compare favorably, of course, with that earned by industrial stocks of a speculative character.

In the United States, as in all young countries where development is rapid, money is at a premium, owing to the strong demand. Consequently, interest rates must be paid that seem extremely high compared with the yields of Old World investments for instance. Whereas, continental investors are content with a return of from $2\frac{1}{2}$ to $3\frac{1}{2}$ per cent with absolute safety, the American investor demands $3\frac{1}{2}$ to $5\frac{1}{2}$ per cent on the same kind of securities. As stated in Chapter VI, it has been clearly proven that a pine plantation, economically started, will yield 5 per cent compound interest, based on present stumpage values. Nevertheless, an interest rate from

Fig. 48.—Sample Plot of White Pine 28 Years Old Which has Been Growing at the Rate of 122 Cubic Feet Per Acre Per Year.

Where initial cost of land and planting is not exorbitant this species will yield 5 per cent compound interest on the investment.

3 to 4 per cent is safer to count on; and the increase in stumpage value that will inevitably occur during the rotation will be an additional profit.

To summarize, under present conditions, forests are a fairly conservative investment, yielding an average return. The legislation favorable to the taxation of forest land that is being enacted by many states, and increased fire protection, are adding to their safety, and natural laws are favorable to appreciation in value. Forestry really offers a unique opportunity to the capitalist, since it constitutes a real estate investment in which the soil receives more protection and less actual labor. Contrary to agriculture, it can be effectively managed by a non-resident owner on account of the extensive character of the work. The crop is not perishable and consequently can be harvested by the owner when market conditions are most desirable.

The small farmer owning a few acres of woodlot also finds his forest land a good investment. The 190,000,000 acres of farm woodlots yielded their owners in 1909, $195,000,000. In addition to their cash value, they supply domestic fuel, building material, furnish labor during the winter season and in many cases, make the farm much more attractive and habitable. From every standpoint, to practically every type of owner from the Federal Government to the small farmer, forest land is a highly desirable investment, one that yields high returns, both direct and indirect.

Methods of Determining Forest Values.

The assessments of any real values is ordinarily quite difficult on account of the many elements that enter into the question. A piece of property may be valued according to its selling value, its cost value (selling value plus interest and expense) or its value may be based on what it is expected to bring at some future time (expectation value) or on its returns (rental value). Sentimental value also must be considered but this is almost impossible to determine by mathematics.

Concerning the first two, little need be said, although there may be wide variation in sale values, due to fictitious conditions or unsound judgment on the part of the vendor.

Expectation value is based entirely on the future sale value which may have to be predicted or estimated; in forest finance it deals with the forest crop. Accurately stated, the expectation value is the exact present net value of a future

FIG. 49. — RESULT OF REPEATED FIRES IN A FARM WOODLOT.

Light ground fires which apparently do no harm beyond burning the dry leaves in reality inflict far-reaching damage which may not appear until the trees are cut.

assumed sale value. In the case of the forest crop, the expectation value is apt to be rather blind because:

1. Exact knowledge of the future stumpage value is lacking.

2. No accurate knowledge of expenses that will be incurred in carrying the crop until time for sale is available. However, it serves to give an idea of the present value when the forest crop is injured or destroyed before maturity.

Rental value is concerned not with the forest crop but considers the forest soil as a revenue producing investment, a capital value which can produce a given return (rental) either annually or periodically. Thus rental value is not concerned with the sale value but rather on the annual or periodic return of the forest; this return (rental) by definition must recur at regular intervals.

In brief the rental value of forest land is its financial value for growing forest crops and in portions of Germany is the value often used for assessment.

Formulæ Used. — In computing the various values and assessing damages, the following symbols are used:

P = per cent, as 3 per cent.
$.0P$ = multiple, as .03.
$1.0P$ = capital and interest for one year.
n = number of years in rotation.
m = number of years to end of rotation.
C = cost of establishing a forest (*e.g.*, planting cost, material and labor).
S = cost of the land.
e = annual per acre expense, taxes, fire protection, etc.
E = capitalized annual per acre expense.

Concerning cost of the land at the end of a rotation of n years, the total expense for land amounts to $S \times 1.0P^n$, but the land still will be on hand and is capable of raising another crop. Consequently its value at that time must be deducted in order to give the correct expense charge incurred in holding the land. If the land can be sold at the end of the rotation for its original price,

$$S \times 1.0P^n - S \text{ or } S\,(1.0P^n - 1)$$

represents the land expense; if the land has changed in value

$$S \times 1.0P^n - S_1$$

is the formula which is universally used, where S_1 equals the value of the land at the end of the rotation. Thus, if the land appreciates in value during the rotation, the total land expense will be less than original price plus interest; if it should depreciate, the total land charge will equal original cost plus interest, plus amount of depreciation; if it remains the same original cost plus interest will be the land expense.

In computing the cost of the annual expense (e), it is much simpler to use the capitalized value (E) in compound interest equations. For instance, if taxes, fire protection and supervision total twenty-four cents per acre per annum with a four per cent interest rate, it will require a capital of \$6 (E) per acre to provide for this annual expense. The total annual expense charge on the rotation (n years) is computed by the formula

$$E\ (1.0P^n - 1).$$

The term -1 is used since the capital sum has never been expended at all, only its interest being used to defray the annual expense.

From the foregoing, the complete cost of raising a crop of timber can be computed by the following formula:

Cost of crop at the end of n years

$$= (C + S + E)\ 1.0P^n - (S_1 + E),$$

S_1 being the land value at the end of the rotation.

Example.

What will it cost to raise a crop of Norway spruce to forty years of age with land costing \$6 per acre, planting cost \$12 per acre, taxes and other annual expenses twenty-four cents per acre; money at 4 per cent, land worth \$10 per acre forty years hence.

Substituting in the cost formula:

Cost of crop at forty years

$$= (C + S + E)\ 1.0P^{n} - (S_1 + E)$$
$$= (\$12 + \$6 + \$6)\ 4.801 - (10 + 6)$$
$$= \$99.22.$$

If the stumpage is worth $100 per acre at forty years, the investment will have been yielding a trifle more than four per cent since the return is slightly more than the interest charged up against the original investment. Any surplus is spoken of as a profit, being the excess over the interest rate charged against the original investment, but the plantation would have been earning 4 per cent had the stumpage yield exactly equalled the cost and no more.

In case a thinning is made previous to the final harvest, the capital plus interest of the financial yield of the thinning is deducted from the cost. The reason for this is that the money received can be put out at interest during the remainder of the rotation and consequently that sum should be a credit charge.

In the above case, if the spruce stand had been thinned at twenty years, the cost would be computed,

$$\text{Cost} = (C + S + E)\ 1.0P^{40} - (S_1 + E + T \times 1.0\overline{P^{20}}) \text{ at 40 years,}$$

T being the net proceeds derived from the thinning.

Assessment of Damages.

Aside from the difficulty in marketing forest products, one of the most discouraging features in the United States has been the slight value placed upon damage to forest lands and standing timber, by the public and by the legal profession. As a result the potential value of immature forest land has depreciated in the eyes of the owners; much of the mismanagement of timberland can be traced to this sentiment.

At the present time the growth of forests has but little standing in the courts and the average judge or jury considers it entirely just and equitable to pay an owner cordwood prices for a burned forest that in a few years would have produced saw timber or at least railroad ties, telephone poles, etc. A woodland owner deprived of future profit is fully entitled to the present value of that profit.

The kinds of damage to forests requiring especial assessment are due as a rule to either fire or trespass, and the damage may consist of the following:

1. Destruction of the merchantable timber.

2. Destruction of future sale value of crop. (Immature crop now unmerchantable.)

3. Injury to producing capacity of forest land or to future forest crops not yet started.

The valuation of these damages may be based on either sale, rental, or expectation value, but there is a wide difference between valuing land for sale and estimating damage to forest property.

If the property remains in the original condition as far as productivity is concerned (trespass case with all timber removed), the stumpage value of the mature timber plus the expectation value of the immature timber will be the measure of damages.

The formula for obtaining the expectation value of the forest crop is:

$$\text{E. V.} = \frac{Y - (S + E)\,(1.0P^m - 1)}{1.0P^m},$$

Y being the cash yield at the end of the rotation; S and E having the same values as before; m being the number of years from the time the damage was inflicted until the end of the normal rotation.

In simplest terms, the expectation value is the final net yield less the future expenses discounted to the present. The original costs including C are not used since they would not represent the value of the crop destroyed except in the case of very young timber. The future expenditures are deducted because they have not been spent and not to deduct them would be equivalent to giving the owner double payment (the full price of his crop which is not yet mature, and in addition the money for expenses not yet incurred).

Example.

The spruce plantation in the previous problem is cut over by a trespasser at thirty years. It would have produced 40,000 board feet worth \$8 per thousand board feet of lumber per acre at sixty years. What is the measure of damages?

The damage being confined to the timber, the expectation value at thirty years will give the amount of damages.

$$\text{E. V.} = \frac{Y - [(S + E)(1.0P^m - 1)]}{1.0P^m}$$

$$= \frac{\$320 - [(\$6 + \$6)(1.4^{30} - 1)]}{1.4^{30}}$$

$$= \frac{\$320 - \$12\,(2.243)}{3.243}$$

$$= \$89.44.$$

In case of partial destruction of a forest crop, the difference between the two yields (normal and actual) $Y + Y_1$ discounted from the end of the rotation to the time of the damage, will give the present value of the loss.

$$\text{E. V. of the original crop} = \frac{Y - (S + E)(1.0P^m - 1)}{1.0P^m}.$$

$$\text{E. V. of the remainder of crop} = \frac{Y^1 - (S + E)(1.0P^m - 1)}{1.0P^m}.$$

The difference between the two is the proper amount of damage or

$$\frac{Y - Y}{1.0P^m}.$$

If the bearing power of the forest land is impaired (land badly burned), its future returns will be smaller and the rental value will be reduced.

The rental value of forest soil, as previously explained, is the capital value based on its regular returns.

By the formula for rental value

$$\text{R. V.} = \frac{Y - [(C \times 1.0P^n) + E\,(1.0P^n - 1)]}{1.0P - 1}$$

$$= \frac{Y - (C \times 1.0P^n)}{1.0P^n - 1} - E.$$

The capitalized value of a series of future net yields is obtained; that is the expenses, $C + E$ with interest to the end of the rotation are deducted from the yield and are divided by $1.0P^n - 1$ to get the *capital value.* The cost value of the land (S plus interest) is not deducted because that would be begging the question. It is the value of the land on a crop-producing basis that is sought.

Example.

The spruce plantation, previously quoted, is entirely burned the first year. Not only trees but all the ground cover was consumed and subsequent rains have washed the soil away. The site is absolutely ruined for timber production.

How much damage shall be sued for?

The soil being impaired, rental value will be the proper assessment, yield Y being secured from yield tables for the same quality of site.

$$\text{R. V.} = \frac{Y - [(C \times 1.0P^{60}) + E\,(1.0P^{60} - 1)]}{1.0P - 1}$$

$$= \frac{\$320 - [12 \times 10.5196 + 6\,(9.5196)]}{9.5196}$$

$$= \$26.35.$$

If the present crop is destroyed and the bearing power of land is also impaired, the usual result of a crown fire, for instance, the rental value of the soil and the expectation value of the crop constitute the two measures of damages.

Presentation of Damages.

The computation of the loss suffered by a forest owner when his young timber is destroyed is comparatively easy; the presentation to a court or jury is extremely difficult as the layman is apt to regard the present sale value as the proper basis for settlement.

The difficulties that lie in the way of convincing the average man of the justice of demanding more than the present sale value are:

1. The growth of the forest is not recognized, the present status of timber being the same as real estate. When the courts are convinced that timber should be considered a growing crop, then it will be possible to substitute expectation for sale value.

2. In the case of damage to young timber, no clear-cut investment is apparent except the purchase of what is usually termed "brush land." If a plantation has been destroyed, it is usually easier to convince the court, as an actual investment for land, planting material and labor has been made.

The following principles should be borne in mind in assessing damages to forest property and in presenting them:

1. Damage must be actual, present or certain to occur.
2. The purpose and use of the owner is considered.
3. The difference in value of the property is the real basis of settlement.

As a rule the simplest way to present an abstruse damage case is to insist that the land of the plaintiff be restored to its original condition. This, of course, is impossible where the soil has been impaired, but by valuing the seedlings that have been injured at $\frac{3}{4}$ cent each when four years old (a plantation of 1200 trees per acre can be easily started for $9) and compounding this initial investment up to the time of the damage, the fact that the present sale value is far too low can generally be established.

Forest Taxation.

A former Chief Forester has said that the two agencies responsible for the greatest amount of forest destruction are forest fires and unjust taxation. Legislation has been passed in some states to improve the situation but where the annual taxes on timberland amount to 6 to 8 per cent of the actual value, a corporation or individual owner is forced to denude the forest land at once in order to prevent taxes from consuming all profit.

It is undoubtedly true that exorbitant taxation has been responsible for much of the forest destruction in the Lake and Pacific Coast States and even now taxes levied upon non-resident timberland owners give them no option but to cut immediately.

Practically all of our states tax forests on the general property basis, and this form of taxation is especially harmful to an investment that does not produce an annual income or one that is either increasing or decreasing in value. Fairchild has shown that a 2 per cent annual tax levied on a plantation of white pine, with land costing $10 and planting $7, would mount up to 78 per cent of the profit at the end of the rotation.

That forest taxation in its present status has not caused more forest destruction is due to the ordinary practice of greatly undervaluing forest land. Where the county officials "cruise" the timber to ascertain the exact contents and then levy an annual tax of 6 to 8 per cent of the value of the stumpage, there can be but one result. One large syndicate of land owners is paying annual taxes amounting to $1,000,000, on forest lands in the state of Washington alone.

The equitable taxation of forest resources is a broad economic problem on the solution of which depends the forest policy of practically four-fifths of the standing timber of the United States. With just taxation many owners can be induced to handle their holdings for a perpetual supply; if the non-agricultural acres containing a splendid stand of timber are taxed exorbitantly, a policy of wasteful exploitation will be put in force, the woodland will be stripped and thrown back upon the state, the counties will be deprived of their revenue and the local inhabitants of their occupation.

The remedy for such a situation is to lighten the burden of forest taxation, to abandon the general property tax now in force in thirty-four states and to substitute a tax on the basis of yield. Moreover, this tax should be collected at the time the owner is best able to meet his obligations; viz., when the crop is harvested.

Fourteen states have enacted special legislation dealing with forest taxation. For the most part, the underlying idea is that of tax exemption or rebates.

Five states have enacted laws permitting the taxation of forest land, according to yield or income. This is the ideal condition, and when every state will permit its non-agricultural acres to pay but one tax, instead of one for each year of the rotation, which tax will be graded according to the producing power of the forest and levied when the owner is in funds, the forest resources of the United States, as a whole, will be handled in a much more rational manner.

BIBLIOGRAPHY.

CHITTENDEN, ALFRED K., and IRION, HARRY. The Taxation of Forest Lands in Wisconsin.

FAIRCHILD, FRED R. Taxation of Timber Lands. Report of National Conservation Commission, Vol. II, page 581.

FERNOW, B. E. Taxation of Woodlands. Forestry Quarterly, Vol. V, No. 4, page 373.

GASKILL, ALFRED. How Shall Forests be Taxed? Proceedings of Society of American Foresters, Vol. I, No. 3, 1905.

SCHLICH, SIR WILLIAM. Manual of Forestry, Vol. III.

REPORT OF FORESTRY COMMITTEE. Fifth National Conservation Congress on Taxation.

CHAPTER XV.

REGIONAL STUDIES.

Forest Regions.

This country was originally endowed with greater and more varied forests than are found in any other country, from a commercial viewpoint. When the earliest settlers came to our eastern coast, there was an unbroken forest stretching from the Atlantic Ocean to the treeless plains of the prairies in the central West.

It is estimated that our original forests covered about 850,000,000 acres. Through clearance for agricultural development, and as a result of logging and forest fires, this amount has been reduced until at the present time there are about 545,000,000 acres classified as forests. It is believed that the present acreage will not be greatly reduced for agricultural purposes, except in certain limited regions, for the reason that the potential possibilities of the soils formerly covered by forests have been nearly reached in agricultural expansion. In many of our states at least one half of the total area is better suited to forests than to any other crop from the soil. In portions of the East soils that should be permanently devoted to forests have in some cases been cleared in the early development of this country and are now reverting back to woodlands and forests.

The demarkation line between the adaptability of all soils to forestry or agriculture will soon be closely drawn in the United States as it has been in European countries, where

ARCTIC OCEAN
GREENLAND
ICELAND
Davis Strait
Hudson Bay
ALASKA
CANADA
Bering Strait
SIBERIA
NEWFOUNDLAND
PACIFIC
OCEAN

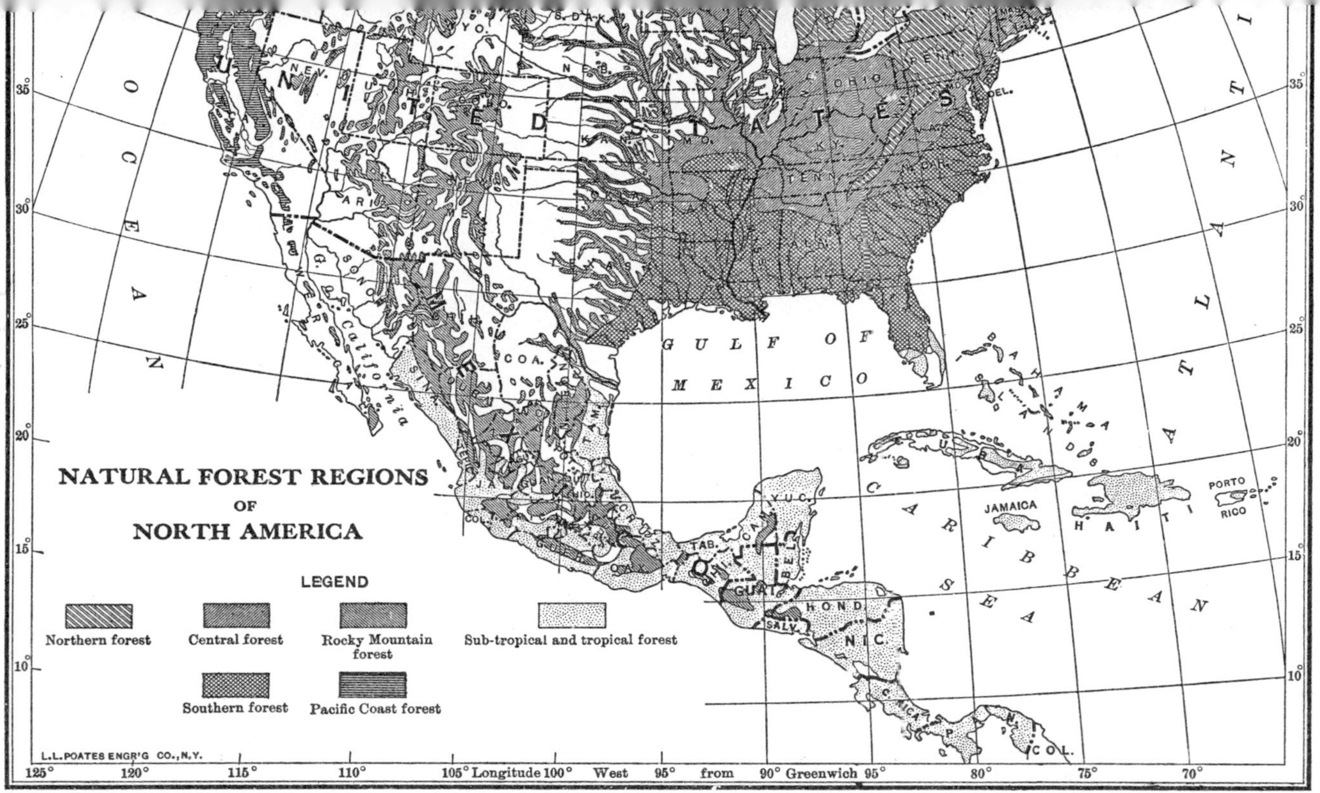
NATURAL FOREST REGIONS
OF
NORTH AMERICA
LEGEND
Northern forest
Central forest
Rocky Mountain forest
Sub-tropical and tropical forest
Southern forest
Pacific Coast forest
GULF OF MEXICO
CARIBBEAN SEA
JAMAICA
HAITI
PORTO RICO
California
L.L.POATES ENGR'G CO.,N.Y.
105° Longitude 100° West 95° from 90° Greenwich

Fig. 50. — A Typical View over a National Forest in the West.

The National Forests furnish both an important source of supply of timber and grazing and serve to regulate the waterflow for irrigation, water power, and municipal reservoir supplies.

every parcel of land is put to its highest economic and productive use. Our oldest lumber producing state, Maine, has the largest percentage of area devoted to forests of any state in the union and there is every likelihood of the permanent continuance of this situation.

The original forests in the East were characterized by the following features:

1. A continuous forest cover, unbroken by treeless plains prairies or deserts.

2. Mixed hardwood and coniferous growth. There are very few pure forests in the East with the exception of the longleaf pine of the South.

3. A great variety of species. There are over 400 species, both hardwoods and conifers, found in the East.

4. Fairly uniform size of trees and stands per acre.

On the other hand the forests of the West are characterized by the following features:

1. Practically pure coniferous forests throughout.

2. The forest cover is usually limited to the mountainous regions and is broken up by treeless plains and valleys and by open parks in the timber.

3. There is great diversity in size, depending on climatic conditions. The large size of the Douglas fir on the Pacific slope is due to the warm, equable, rainy weather, whereas the same tree in the Rocky Mountains is comparatively short, tapered and of small size, due to the low rainfall, short growing season and low mean temperature.

4. There are fewer species than in the East. There are only about eight important lumber trees in the West.

The principal species, from a commercial standpoint, found in the East are white pine, longleaf pine, white and red oaks, hard maple, shortleaf pine, cypress, red spruce, hemlock,

FIG. 51. — PANORAMA IN THE SOUTHERN APPALACHIANS, YANCEY CO., NORTH CAROLINA.

The government has provided for the purchase of National Forests in the East. About 1,000,000 acres are being reserved for this purpose in the southern Appalachian Mountains and the White Mountains of New Hampshire.

loblolly pine, red gum and yellow poplar. Besides these is a great variety of species, such as the chestnut, red pine, elm, hickory, eastern larch, white and red cedar, black gum, ash, walnut, cherry, basswood, etc.

The principal species in the West are Douglas fir, western yellow pine, redwood, western red cedar, sugar pine, lodgepole pine, western hemlock, Sitka spruce and western white pine. Others of less importance are western larch, incense cedar, red fir, Port Orford cedar, amabilis fir, Engelmann spruce and a few other conifers. The few hardwoods on the Pacific slope, among which may be mentioned the California live oak, the tan-bark oak and the Oregon maple are of relatively little importance.

About 50 separate species enter into our timber trade in an important way. About 150 others are used to some extent but chiefly for local purposes. In keeping with our rapidly increasing population, and the gradual western development, the supply of white pine, our best all around timber tree, has rapidly diminished. Southern yellow pine, embracing four species (longleaf, shortleaf, loblolly and Cuban pines in order of importance) was next exploited and at the present time, Douglas fir on the Pacific coast is rapidly coming to the front to supply our lumber and timber for the various wood-using industries.

The result of this situation is that the eastern forests have been culled over for the best material and the remaining supply is being cut from the more inaccessible portions of the virgin forest. This means that forest management will undoubtedly be practiced first in those portions that have been longest settled and therefore have suffered most from the result of the axe, forest fires and clearing land for agricultural development. For example, New England has taken the leadership in the practice of forestry for the reason that it is one of the oldest as well as most densely populated sections of the country. In this region both the needs and possibilities of forestry are closely comparable to those of Germany and France. In the Northeast the most direct reason for the great interest in forestry is the question of utilization. In other words, forest products bring such attractive prices that intensive forest management is possible and practicable.

For the purpose of briefly describing the forestry situation in the various parts of this country, the forests have been divided up into several regions. In the ensuing chapters the following regions are described in a general way:

(1) Northern Forest,
(2) Southern Pines,
(3) Central Hardwoods,
(4) Prairie or Fringe Forest,
(5) Northern Rocky Mountain Forest,
(6) Southern Rocky Mountain Forest,
(7) Pacific Coast Forest.

These regions are divided according to differences in species and prevailing types of timber which serve to distinguish them. A brief description of the most important phases of the forestry situation in each region is taken up. For the purpose of brevity, these regions have been subdivided along broad lines. For example, in the central hardwood region there are included the following local regions; the Piedmont Plateau, the eastern sprout hardwoods, the central woodlot region, etc. In some cases, as in the northern and southern Rockies, similar types of timber occur in two different regions so that the lines of demarkation between the regions are very broad and general.

BIBLIOGRAPHY.

CLEVELAND, TREADWELL, JR. — The Status for Forestry in the United States. Forest Service, Circular 167.

FOREST SERVICE. — Circular 171. Forests of the United States: Their Use.

ZON, RAPHAEL. — The Future Use of Land in the United States. U. S. Forest Service, Circular 159.

ZON, RAPHAEL. — Forest Resources of the World. U. S. Forest Service, Bulletin 83.

CHAPTER XVI.

NORTHERN FOREST.

Location and Boundary. — The portion of the timbered area of the United States considered as the Northern Forest comprises the northern portion of Minnesota, Wisconsin and Michigan, and ranges from there south along the Appalachian Mountains as far as northern Georgia. In the Northeast the bulk of Pennsylvania and New York and all of Vermont, New Hampshire and Maine are included within this region.

The total area of land surface, as included within these limits, will approximate 240,000,000 acres, gross. The net area, or the portion within this region actually covered with forest, would be extremely difficult to estimate. Maine, for instance, is very heavily wooded, having approximately two-thirds of its surface heavily covered with forest, while other states, like Massachusetts, have a comparatively small portion covered with forest, and still others, like Michigan and Minnesota, which were primarily lumber states and possessed at one time splendid areas, now have their timbered areas badly broken by vast stretches of waste land, which are the results of careless cuttings and forest fires.

The territory which is here described in reality contains a wide variety of forest types and conditions. The changes in

Fig. 52. — Outlook from Fire Station in the Adirondacks.

The view is typical of the cut-over spruce regions in northern New York. If forest fires follow lumbering, the litter may be consumed and nothing left but bare rock.

forest cover are largely due to topography and climate, and so a word or two concerning the physiography of the region is necessary.

Practically all of the northern portion was covered by the ice pack during the glacial epoch, and as a consequence, the soil, topography and drainage are decidedly irregular. Vast outwash plains of nearly pure sand may alternate with heavier clays; drumlins in great number are found, and countless lakes, swamps and slow streams are also found, especially in northern Maine and New York.

The southern part, including the Appalachian Mountains, has never been glaciated, consequently the drainage is more regular and the numerous swamps are lacking. The soils, however, are quite variable, ranging from the barren sand to heavy clay loams. The sandy soils, however, are much more frequent.

The topography presents many striking variations from the sandy flats of the Lake States to the rugged mountain slopes of the White Mountains and the Appalachians. While there are considerable stretches of fertile agricultural land, especially in New York and Pennsylvania, rugged topography is the rule. The maximum elevation is reached in North Carolina (Mt. Mitchell, 6711 feet), with Mt. Washington (New Hampshire), 6290 feet, a close rival. The majority of this region lies above 500 feet, and most of the southern extension above 2000 feet. The average annual precipitation ranges from twenty to seventy inches, the latter being the maximum in the southern Appalachian region.

Forest Characteristics. — Within this region several subtypes or subregions might be differentiated, viz., the spruce region of the Northeast, the pine region of the Lake States, and the hardwood region of New York, Pennsylvania and the Appalachian Mountains. The characteristics of each of these will be briefly described.

The spruce forest of New England and northern New York is confined principally to the higher slopes of the mountains. It is composed in the main of tolerant species which form a mixed forest of uneven ages. The chief species are red, white and black spruce, red and white pine, balsam fir, hemlock, tamarack, white cedar, hard maple, beech and birch (yellow and paper) and aspens are found in considerable quantity. The optimum region for red spruce occurs within the area. On the headwaters of the Androscoggin River and upon the upper slopes of the White Mountains the heaviest stands of pure spruce are found. The spruce stands of the Adirondacks, on account of the gentler slopes and deeper soils, are apt to have hardwoods intermixed to the extent of fifty to sixty per cent.

The hardwood part of this region contains two distinct subdivisions, the so-called northern hardwoods, and the Appalachian hardwoods.

The former is found in the New England states and in New York at a lower elevation than the spruce type; it is composed of the same shade-bearing hardwoods found mixed with the spruce, viz., beech, yellow and paper birches, and hard maple in the virgin forest, while the aspens are found to a large degree taking possession of the burns.

The southern portion of this subtype is found at higher elevations than the former; the species are quite different, being intolerant for the most part, and the conifers present are found only at the highest elevations. The abundant rainfall, the long growing season and the soils, for the most part entirely adequate, make this an ideal region for tree growth. Out of the 500 tree species indigenous to the United States, approximately 135 are found in the Southern Appalachians. Chestnut, basswood, chestnut oak and yellow poplar find here the optimum region for their development, while such species as black, white and red oak are found of splendid size and development, but not in their optimum region.

Perhaps the best example of a true forest region may be found in the Lake States portion of the northern forest. With extremely light soils the rule, with fairly high precipitation ranging from twenty to forty inches per annum, with vast areas of swamps, muskegs and barrens, a large portion of this subdivision is better suited to raising timber than to agriculture. Here the forest is largely coniferous, and occurs for the most part in even aged stands. White pine is found at its optimum, with unusually favorable conditions for Norway and Jack pine. Of the hardwoods found, yellow birch, hard maple and beech are the most important. Elm, basswood and paper birch are found mixed with the other hardwoods on the heavier soils, but only in light stands. White pine is found on the lighter soils and may be intermixed with aspen and paper birch on moist sites, and with Norway pine on the drier ones.

The lightest and driest soils are generally occupied by Jack pine, a species which shows a remarkable power to occupy areas laid waste by fire. Considering the forest as a whole, we find lumbering the oldest industry in the northern forest. The first sawmill in the New World was erected at Berwick, Maine, in 1631. While sadly diminished by reckless cutting followed by forest fires of which the Lake States pineries can furnish many mournful examples, this region is still supplying considerable timber and pulpwood. With proper management it will continue to do so for some time to come.

Silvicultural Treatment. — On account of the variety of conditions found within this region no single method of forest management can be prescribed.

In the spruce region selection cuttings are generally practiced, although on shallow soils and steep slopes clear cuttings may be necessary. When mixed forests are handled, the hardwoods are generally cut to as small a diameter limit as the market will permit, in order to favor the faster growing and more desirable spruce.

The recent increase in the value of the northern hardwoods has made more intensive forestry practice possible. One of the largest paper concerns of the Northeast has been thinning its hardwood land quite heavily, and underplanting with 300 to 500 Norway spruce seedlings per acre.

Within the pine forests in the Lakes States and to a less degree in New York and Pennsylvania, a clear cutting system must be used as their light requirements will not permit the use of the selection system. The ground may be cut entirely clear, leaving three to five seed trees per acre, or young thrifty trees may be left in groups, twenty to thirty per acre, which will eventually cover the area with young seedlings. Leaving seed trees in groups is to be preferred in regions where high winds prevail, or where the trees to be left are not especially windfirm.

In the northern hardwood portion, extensive management will be practiced for some time, in spite of the fact that the forest occurs largely in connection with agricultural land. The trees being tolerant, they, for the most part, can be handled according to the selection system, and the undesirable trees removed to favor such species as maple, birch, ash and basswood. In the case of the two latter species, more light may be needed than the selection system will give, and the group stand method or shelterwood system may then be used.

The Appalachian hardwoods present a rather difficult problem in management, on account of the extreme complexity of the forest. Such species as cucumber, white ash, yellow poplar and white oak should be favored, but, as a rule, they have been largely removed from the stand during previous lumbering operations. The problem consists of perpetuating the trees of greatest market value; those which it is the greatest temptation to cut. Clear cutting with proper seed trees is the best system to employ, and in some cases merchantable trees must be left in order to provide for seed. Undesirable species should be cut clear if a market is available; if not, "girdling" may be practiced, which will prevent them from reproducing or shading out more desirable reproduction.

Protection. — Throughout the coniferous portion of the Northern Forest, fires by all odds inflict the greatest damage. Vast areas have been laid waste in the pineries of the Lake States, and the steep slopes of the White Mountains and Adirondacks have been heavily cut and badly burned. Where the thin soils of mountainous regions are burned over and then exposed to the wash of melting snows, the destruction is seen at its worst. Practically nothing but rock is left, and it will take centuries of weathering to produce enough soil to grow a merchantable crop of timber.

The responsibility for fires may be divided among railroads, sportsmen, farmers and ignorant foreigners from the manufacturing towns. Legislation controlling the railroads has been passed in practically all of the northern states. Spark arresters are required, ash pans must be kept in repair, and in New York state, the Public Service Commission has forced the Adirondack railroads to burn oil in their locomotives during the danger period.

Controlling a responsible corporation is a comparatively easy task when the aid of a State Commission is invoked. In addition the transportation companies are beginning to realize that attractive green forests are an asset both from an esthetic and a commercial standpoint. The irresponsible individual, who sets the woods on fire with a pipe, camp fire, or when burning fallow, is much more difficult to reach. Education of the public, commencing with the school children and carried on for years, is the ultimate solution. This method is being pursued by the State Foresters and educational institutions within this region. Protective machinery to prevent and combat fires, forest rangers, lookout stations, etc., are very necessary, but in the final analysis, it is a matter of educating the people. Certain precautions to prevent fires are possible, and brush in extremely dangerous localities should be gathered in piles or burned. The New York Conservation Commission has recently reinforced the toplopping law, after a successful trial extending over three years.

Other injurious agencies that might be mentioned are windthrow and windbreak, occurring quite frequently in stands of spruce and balsam. If light thinnings are made on ordinary sites, and if clear cuttings are made on exposed localities, serious damage may be avoided.

Insects are occasionally troublesome. The spruce bark beetle has killed considerable timber at different times, especially in the Androscoggin drainage. The white pine weevil attacks the leaders of young, open-grown pines in various sections, and the larch sawfly has on several occasions badly damaged the northern forests.

By cutting and soaking the infected spruce timber after August, the beetles may be disposed of; the weevil can be checked by removing and burning the infested tops during June and July; the sawfly practically defies attack, but since 1882 the amount of damage inflicted has been quite restricted.

In the parts of this region where the northern hardwoods prevail, fire damage is much less serious on account of the less inflammable nature of the stand. The leaves of the hardwoods decompose more readily and the forest floor as a consequence is less inflammable. In addition, the broken nature of the woodlot renders extensive fires impossible.

Grazing, however, is particularly harmful in northern hardwood forests. Cattle are especially fond of browsing on the seedlings of sugar maple, and, as a consequence, are apt to interfere with the natural regeneration of this desirable species. No woodlot which contains young reproduction or which is to be regenerated by natural means should ever be grazed. Seedlings will be bitten and trampled, and the soil rendered too compact for a tiny seedling to take root.

In the Appalachian country grazing is still worse than in the North. The cattle browse upon all the succulent hardwood seedlings and are especially fond of tulip poplar; the hogs with their keen appetite for acorns practically prevent the seedling regeneration of the oak. The most important feature of grazing, however, is its bearing upon forest fires. The inhabitants of this region believe that frequent forest fires improve the pasturage and unless stock laws can be passed which will restrict cattle and hogs, annual forest fires will continue.

Watershed protection is a phase of forest management that here assumes enormous importance. With the steep slopes and compact soil that prevail and the heavy rainfall descending in frequent thunder storms, erosion and flood inflict a heavy toll. Steep slopes have been cleared which should have remained in forest. Heavy grazing and forest fires have added to the predisposing influences. The damage is not only inflicted where the slopes have been denuded by the flood waters, but they submerge fertile bottom lands in the Piedmont region and often render them useless by covering them with layers of sand and gravel. Extensive reforestation of the steeper slopes as a reinforcement to a series of dams and reservoirs, is the only practicable solution of this problem.

The Federal Government, acting under the Weeks Act, has already acquired about 1,000,000 acres in this region confined entirely to the headwaters of navigable streams.

Utilization. — The manufacturing and marketing of the products of the northern forest is a problem comparatively easy to solve on account of the dense population of the region, thereby creating a strong demand and a stable market. The transportation facilities also are entirely adequate to a proper distribution of both raw and manufactured material.

Within the spruce region the situation is relatively satisfactory. The countless streams permit economical transportation of logs to the manufacturing centers and the fact that there is hardly a point in the region more than 300 miles from a city of the first or second class makes for profitable marketing of forest products.

Wood pulp is the principal product, and two states, New York and Maine, together consume about 2,000,000 cords per annum. This has proved to be such a profitable form in which to market the soft woods that some of the largest stumpage owners have discontinued the manufacture of lumber and are putting all of their material into pulp.

The northern hardwoods are at present rapidly appreciating in value, and their uses are increasing in number. Aside from furniture, flooring, etc., minor industries, like last blocks, basket veneers, slack cooperage, whip butts, etc., are examples of the forms in which the hardwoods finally appear upon the market. Paper birch has especial value for shoe pegs and spools, which are turned out in large quantities.

The pine lumber of the East is largely manufactured into box and barn boards, bucket stock and interior trim. On account of the variation in sizes needed, very close utilization is possible. In fact, in certain parts of Massachusetts, white pine may be cut to a three-inch top. The Lake States pine is cut and manufactured on a far more extensive scale and markets may be a considerable distance from the raw material. Lumber of all dimensions is the product, and the utilization is by no means as close as in the East.

FIG. 53. — VIRGIN FOREST OF WHITE PINE, MINNESOTA NATIONAL FOREST, MINNESOTA.

First growth of white pine may become 6 feet in diameter and 160 feet high. Such specimens take from 250 to 300 years to attain this size.

The hardwoods of the Appalachian region have had a rising market for the last two decades, but the end of virgin timber is already in sight. The most important local industry has been lumbering, with railroad ties as a very important phase. Within the last few years the number of portable sawmills has greatly increased and this fact, in connection with the opening of new areas of soft coal, will permit of more local manufacture, and a closer utilization of hardwoods for mine props, logging, etc. Minor industries, like the manufacture of cooperage stock, and the gathering of tan bark, etc., are quite common. Chestnut extract has been of great importance on account of the larger percentage of tannin in southern than in northern hardwoods. With the spread of the chestnut blight this region may be forced to utilize enormous quantities of chestnut wood in this fashion.

Special Problems. — From the standpoint of Federal importance, the Appalachian Acquisition in compliance with the Weeks Act is doubtless the largest problem in the region. For years the farmers in the lowlands have had their lands submerged by disastrous floods, due primarily to the combination of overcutting and forest fires on the steep slopes of the headwaters.

Owing to the fact that these headwaters were often in other states, Federal aid was necessary, and in 1910 the Weeks Act was passed by Congress to permit the government to acquire land upon the watersheds of navigable streams. Already, over 1,000,000 acres have been purchased, largely in the southern Appalachian Mountains.

It is planned to create National Forests in these regions and to so manage this land as to improve its water-holding capacity, thereby greatly reducing the damage formerly occasioned by erosion and floods.

In the Lake States planting the enormous stretches of

barrens and forest land laid waste by fire is a problem of great importance, with taxation and fire risk as limiting factors. In addition, a mistaken conception of the value of some of these lands for agriculture has checked the proper development of a forest planting policy.

In the Northeast, notably in Maine and New York, the proper development of water powers is of increasing importance. With discoveries permitting long-distance transmission of electric power the value of mountain streams and lakes is rapidly appreciating and the importance of forests in controlling run-off in some cases equals their value as sources of timber.

Future of Forestry in this Region. — It is safe to say that no forest region in the United States has a brighter outlook for the practice of intensive forestry during the next two or three decades than the northern forest. With the center of population but a few hundred miles west of its margin, with splendid markets and transportation facilities already developed, with vast areas better suited to timber production than to agriculture, the conditions necessary for extensive forest management are even now satisfied. First to be exploited by the colonists, the recuperative power of the northern forest has proved remarkable. The Lake States, considered exhausted twenty years ago, are still yielding splendid harvests of timber; Maine cut over first for pine masts is now yielding her third and fourth crop of timber.

With proper protection and intelligent use, the forest soils of this region will supply its inhabitants with timber, building materials, etc., for centuries to come.

BIBLIOGRAPHY.

BOWMAN, ISAIAH. Forest Physiography. John Wiley and Sons, New York.

CARY, AUSTIN. Practical Forestry on a Spruce Tract in Maine, U. S. Forest Service, Circular 131.

Graves, Henry S. Practical Forestry in the Adirondacks. U. S. Forest Service Bulletin 26.

Hawley and Hawes. Forestry in New England. John Wiley and Sons, New York.

Pinchot, Gifford. The Adirondack Spruce.

Roth, Filibert. Timber Conditions in Minnesota, Twenty-first annual report, United States Geological Survey, page 6.

Zon, Raphael. Result of Cuttings on the Minnesota National Forest. Proceedings of Society of American Foresters. Vol. VII, No. 1.

CHAPTER XVII.

SOUTHERN PINES.

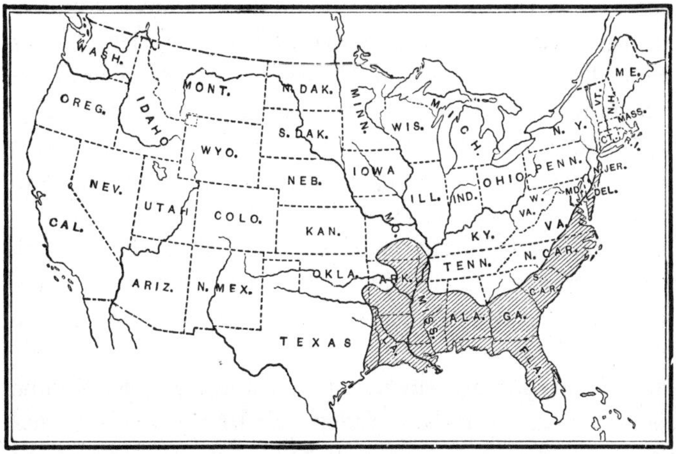

Location and Boundary. — The southern pine region includes the portion of the country along the Atlantic and Gulf coasts from southern New Jersey to eastern Texas. For the greater part it is a belt of forest stretching from 150 to 250 miles in from the coast, although it begins as a narrow point at the extreme north and gradually widens to its greatest extent in the lower Mississippi River valley. For the most part it covers the coastal plain of the South.

The region is largely bounded on the north by the central hardwood region. On the west it touches the prairie or fringe forest. It includes all of Louisiana, which is the principal center of yellow pine production, and most of North Carolina, South Carolina, Georgia, Florida, Alabama, Mississippi, Arkansas and eastern Texas.

For the past decade the southern pineries have been the producing center of our lumber supply and at the present time, the four southern yellow pines, longleaf, shortleaf, loblolly and Cuban, comprise over one third of the total annual lumber cut of the United States. This region also produces all of the turpentine, rosin and other naval stores.

The topography is generally flat or gently rolling. Along the foothills of the Piedmont Plateau, however, some of the land included within the southern pineries is distinctly hilly, as in Alabama. Throughout the region the broad plains or flats are broken by low lying stream courses which are often estuaries for several miles from the coast.

The climate may be characterized as warm and humid with abundant rainfall, all of which are very conducive to tree growth. The growing season is the longest of any region excepting in California.

The soils are largely sandy, especially in Georgia, Florida, Texas and South Carolina, giving rise to the name "sand barrens." These are largely responsible for the slow growth of the pines.

The southern portion of the Florida peninsula contains a distinctly tropical vegetation and is therefore not included in this region. It is characterized by a heavy precipitation, 60 to 70 inches, and the presence of a great variety of species, about 75 of which are not found in other parts of the country. The principal species are the mangroves, Jamaica dogwood, mahogany, lignum vitæ, corkwood and the mastic. The most valuable trees are the mangroves, on account of their soil-forming qualities along the sea coast and the fact that they contain a high per cent of tannin.

Forest Characteristics. — This region is largely coniferous as opposed to the other eastern regions and the forests cover about 75 per cent of the total land surface. It is estimated that between 40 and 50 per cent of the area will always be retained in forest growth, on account of the sterility of the sandy pine barrens for agricultural purposes.

FIG. 54. — TURPENTINE ORCHARD, SANTA ROSA CO., FLORIDA.
The new system of collecting the resin in cups is less wasteful than the old box system.

This is the home of the longleaf pine, the most important tree of the region, and which, together with Douglas fir, produces a good share of the country's timber supply.

The forest in the Southeast may be characterized as follows:

(1) Preponderance of coniferous growth.

(2) The present producing center of the country's lumber supply and therefore a very active logging region.

(3) The forest is severely affected and injured by surface fires.

(4) The forest is being severely injured by the old methods of turpentining.

The principal conifers in order are: longleaf pine, shortleaf pine, loblolly pine, cypress, Cuban pine and red cedar. All of these, excepting cypress, grow on the drier uplands and sand barrens. Loblolly pine and Cuban pine also require some moisture.

The hardwoods grow along the streams and in the bottomlands and swamps in association with the cypress. Most of the latter is limited to the swamps bordering the coast and the lower Mississippi Valley. The principal hardwoods are tupelo, cottonwood, red gum, ash, yellow poplar, sycamore, elm, several oaks, hickories and hackberry.

The forest may therefore be divided into two broad types based on site factors, that is, the presence of moisture, character of soils, etc.

In the coniferous type there are several variations, ranging from the pine longleaf or pure shortleaf type to the mixed conifers and hardwoods and the loblolly pine type. The Cuban pine requires the most moisture of any of the pines; loblolly pine requires at least a moist, deep soil; but longleaf seems to flourish on dry sandy barrens where no other trees seem to grow. The latter is rarely found on moist soils. The average coniferous forest will cut about 5000 board feet to the acre although many stands will average from 10,000 to 20,000 board feet in restricted localities.

The hardwood and cypress type changes with the relative amounts of moisture present. In the bottomlands, overflowed a portion of the year, cypress sometimes grows in pure stands. Usually, however, water gums, water hickory and red bay are associated with it. Along most of the streams, hardwoods predominate to the exclusion of the cypress. Altogether in this lowland type, stands running up to 50,000 board feet to the acre are common. The average merchantable stand will cut from 8000 to 12,000 board feet to the acre.

Silvicultural Treatment.— No systematic plans for forest management have been adopted for any length of time in this region so that it has had very little silvicultural treatment of any kind. Practically all of the forest is in the hands of large owners instead of at least partial state or Federal government control as in other regions. The forest is being rapidly cut off and on account of the prevalence of fires, reproduction has a poor chance to get started.

The longleaf pine grows in unevenaged stands or groups and will be handled either by a rough adaptation of the shelterwood system or a clear cutting and leaving seed trees. It grows exceedingly slow and therefore the more rapid growing loblolly and shortleaf pines should be encouraged in its place wherever possible. Because of its slow growth, longleaf pine as well as cypress, for the same reason, will probably not be important trees in the future. They are certainly too slow growing for forestry purposes. The most hopeful tree from the viewpoint of the future is the loblolly pine. It seeds abundantly, reproduces thriftily and vigorously and is one of the most rapid growing trees in the country under normal conditions. Its wood is not of such high technical qualities as the longleaf or shortleaf pine but it yields a wood that is very good for general lumber and boxboard purposes.

The hardwood type will be handled on some adaptation of the selection system. Most of the trees grow fairly rapidly and many of them will flourish under shade so that the larger specimens can be cut out from time to time as they reach or approach maturity. The best trees to favor in the bottomlands and along the streams are ash, oaks, yellow poplar, cottonwood, red gum and hickory on account of their high commercial value and comparatively rapid growth.

Very little planting has been resorted to in this region. The lumbered areas have either been cleared for farming or

Fig. 55. — A Cypress Swamp along the Lower Mississippi River.

Cypress is one of our most valuable timber trees on account of its durability, soft, even texture, and excellent seasoning qualities. Owing to the difficulty in reproducing itself, it will not be an important tree of the future.

left to burn over annually. The best trees for planting are loblolly pine, red oak, yellow poplar, ash and cottonwood. The latter is easily reproduced from cuttings.

Protection. — The South undoubtedly suffers as much from forest fires as any other region. They are limited, however, to grass fires. But these burn over the forest practically every year and great damage results not only in the prevention of satisfactory reproduction but in the burning of merchantable timber and the destruction of the turpentine boxes.

The open character of the timber growth permits of a loose, grassy undergrowth. In the spring and fall the dried grass is usually fired by the natives to provide better forage for the cattle and hogs in the summer season. This has been such a deep-rooted custom that it is very difficult to advocate any protective measures.

Fires are especially disastrous to the turpentine industry. Fires get into the boxes and the base of the trunk and the tree is eaten away or is easily blown over by the wind.

Fires can be easily stopped by plowing in the porous sand; by beating them out; or by stopping them at roads, trails, fences, etc.

Insect depredations have been spasmodic in the South. The southern pine beetle has broken out in several recurrent attacks, doing enormous damage to standing trees.

Fungi are most abundant on account of the humid, moist climate, and much of the standing timber is heart rotten. "Bluing," a stain on the sap of freshly sawed lumber, is also a common result of fungous action in this region.

Utilization. — As mentioned before, this is the producing center of both the lumber and naval stores industries and the vast pine stands have been so abundant that little attention has been paid to a close and economic utilization of the raw products. However, with the rapid logging of the forests, improvements in the tapping of the longleaf pine for turpentine and rosin have been introduced by Dr. Herty (see Chapter X on the naval stores industry) and others with a considerable saving. Within recent years great improvements have been made in the utilization of the refuse of the large sawmills chiefly through pine distillation. Most of the lumber from this region is consumed in the North and altogether the opportunities for close utilization outside of wood distillation and the naval stores industry are somewhat limited. Some of the large sawmills are now utilizing their waste in the manufacture of paper.

At the present rate of cutting the South will soon be depleted of her virgin timber supplies unless proper measures are adopted to perpetuate the forests. The lumber industry is one of the greatest assets of the South and it could be placed on the same basis in respect to its forests as the other regions if a better public spirit in regard to its natural resources were aroused.

Special Problems. — The special problems peculiar to this region may be summed up as follows:

(1) The prevention of grass fires. The solution of this problem will materially result in perpetuating both the timber supply and the naval stores industry. Fires are not necessary for good forage for the cattle and hogs and the lumbermen are rapidly becoming converted to the opinion that fires are very detrimental to their interests.

(2) The awakening of the public opinion on the forestry situation. The South has been the most backward in regard to both state and private activity in conserving its forest resources. Public concern in regard to the prevalent forest fires is still to be aroused. With the further economic development of the South and the rapid cutting of the yellow pine and cypress stands, the public is sure to take the same interest in forestry as in New England and the Middle Atlantic States.

Future of Forestry in Region. — As stated before about 40 to 50 per cent of the southern pine region will probably always be devoted to some form of forest growth. Sooner or later fire protection must be maintained on this area, which will be the first step in the progress of forestry in this section. With fire protection assured, the forests will probably adapt themselves to the various forms of forestry practice.

Longleaf pine and cypress will probably be eliminated in the future forests and other rapid growing trees will take their places.

The future of forestry is assured in this region wherever the fire problem is solved for the following reasons:

(1) Rapid growth, caused by the warm climate.

(2) Easy reproduction, providing fires are kept out.

(3) Easy logging and log transportation conditions.

(4) There are large areas unsuited to agriculture that must be devoted to forest culture.

BIBLIOGRAPHY.

CHAPMAN, C. S. A Working Plan for Forest Lands in Berkeley Co., South Carolina. U. S. Forest Service, Bulletin 56.

FOSTER, J. H. Forest Conditions in Louisiana. U. S. Forest Service, Bulletin 114.

MOHR, CHAS. Timber Pines of the United States. U. S. Forest Service, Bulletin 13.

REED, F. W. A Working Plan for Forest Lands in Central Alabama. U. S. Forest Service, Bulletin 68.

SCHWARTZ, G. F. The Longleaf Pine in Virgin Forest.

ZON, RAPHAEL. The Loblolly Pine in Eastern Texas. U. S. Forest Service, Bulletin 64.

CHAPTER XVIII.

CENTRAL HARDWOODS.

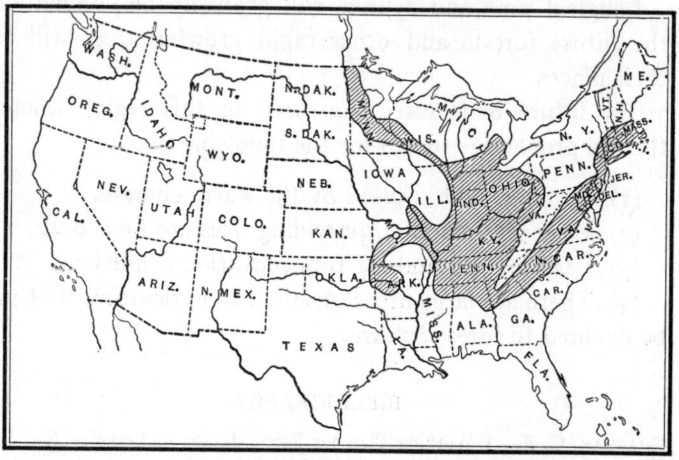

Location of Boundary. — The forest area that is described as the central hardwoods region extends from Massachusetts on the north, south along the Piedmont plateau between the Appalachian Mountains and the Coastal Plain to northern Georgia and Alabama; northwest through central Illinois to western Minnesota. Its area contains approximately 250,000,000 acres.

The topography is comparatively low and rolling, the elevation ranging from sea level to one thousand feet. The soils are deep, and often rich enough for agriculture. The small amount of swamp land present can often be drained, so that there is little true forest soil within the region.

The annual precipitation ranges from thirty to forty-two inches, and is well divided between the dormant and growing seasons. With soils of agricultural character prevailing, and with abundant precipitation, this region is destined for agricultural development, and will never produce timber enough for export. However, fencing, fuel, and domestic building material must be raised, and in some portions the dual purpose windbreak is of great importance.

Small holdings are the rule in the central hardwoods region, although in the southern parts some good-sized timber tracts may be found. The bulk of the timber, however, is in the form of farm woodlots, or more precisely stated, is owned in connection with agricultural land.

Forest Characteristics. -- The forest is characterized by the following distinctive features:

1. The predominance of hardwoods. The old field pine type of the Piedmont regions is not to be overlooked, but its importance is relatively small in comparison with the hardwood type. A large portion of the hardwood timber supply in this country, constituting about one-fifth of the total annual lumber cut, comes from this region.

2. Large number of trees found in mixture. Within these boundaries are found the regions of optimum development of such species as white, red, black and pin oaks, yellow poplar, and beech, while hickory, white ash, black walnut, cottonwood, maples and cedar are also present in considerable quantities. In Illinois alone, there are found one hundred native tree species out of the five hundred common to the United States.

3. Broken distribution of the woodland. This condition obtains to such a marked degree, especially in the western central part of the region, that the management of the forest

is, to a large degree, affected by the absence of continuity. Wherever rough topography or low lands along rivers are found, the timber may occur in large bodies.

The stand, as a whole, may be characterized as second growth hardwoods, containing a considerable number of veterans, singly or in groups, left over from the first cutting. The present condition of the woodland proves conclusively that a forest region may deteriorate as much when poorly managed as when wilfully abused, for the chief trouble has been the wrong point of view. There has been no extensive lumbering followed by forest fires which laid bare enormous areas as in the lake states, but the forests have not been regarded as an asset. For the most part, they have been considered an encumbrance to be removed with all speed, and, in some cases, timber has been piled and burned to clear the ground for tilling. In many cases, slopes have been clear cut that should have remained under forest cover, and now, rapidly eroding hillsides and gravel-covered bottom lands prove conclusively that such areas should have been kept in forest.

Where clear cutting has not been practiced, the forest, for the most part, has been extensively culled. The largest and best trees have been removed, leaving the less desirable specimens and forest weeds in possession of the soil. Thus, each successive tree generation has become less valuable than the previous one. Fires have inflicted the worst damage along the foothills of the Appalachians, but throughout the bulk of the region they can be quite easily controlled on account of the broken character of the woodland.

Grazing is responsible for much deterioration of the forest in states like Ohio and Indiana. There, the woodlot is largely considered a part of the pasture with the result that it serves neither purpose well. The trees shade out the grass to a large

FIG. 56. — SOUTHERN APPALACHIAN HARDWOODS.
Chestnut, oak, ash, and yellow poplar in western North Carolina.

extent, and the open forest floor, windswept and sunbeaten, is not sufficient to grow timber at a profitable rate.

Silvicultural Treatment. — The most important step in improving forest conditions within this region is to change the point of view. As soon as the value of forests and wood products is realized, the owners themselves will take better care of their non-agricultural areas, and will endeavor to make them profitable. Much propagandist work has already been done by the Federal and State Forest officers that will eventually show results. The actual plan of management in this region will vary considerably, depending on the type of forest, nearness to market, and relative intensiveness of management.

In the Northeast, where sprout hardwoods prevail, the coppice system of management will doubtless be employed, with chestnut badly handicapped on account of the fungous disease (Endothea parasitica). The mixed forest should be the ideal, on account of its greater productiveness, and the reduced danger from fungous and moth attacks.

Along the Piedmont area, wherever the shortleaf and loblolly pines form any part of the stand, their growth should be encouraged. The less desirable hardwoods should be discriminated against by heavy cutting, and the soil may, in some cases, be prepared by running hogs through the woods just before the fall of seed. Their sharp hoofs will break up the earth, and make a good catch more probable.

In Ohio and Indiana, artificial regeneration must be resorted to on a large scale. The woodlots, for the most part, are open groves with very compact soil. The leaves have blown away for so many years that there is very little humus in the soil; the remaining trees are generally of the inferior varieties, or diseased specimens of the desired species. On the whole, better results will be obtained if such woodlots are clear cut and replanted, thereby obtaining the desired mixture and spacing at once.

In bottom lands, where only a light culling has been made, natural regeneration is possible, but this, as a rule, requires a good broad market that will take care of all sizes and kinds of lumber.

As before stated, planting in the open for windbreaks which furnish both fuel and protection is highly desirable wherever drying winds prevail.

The species to favor where natural regeneration is employed, would be as follows:

In the Northeast: White and red pines, red oak, ash, tulip, poplar and basswood.

In the Piedmont area: Ash and tulip, red and white oak, hickory, loblolly and shortleaf pines.

In Ohio and Illinois: Ash, yellow poplar, cucumber, black locust, red and white oaks, black cherry and hickory.

In the latter region, white pine, white cedar, elm, red maple, ash, willow and the Russian wild olive are desirable for windbreak purposes, while catalpa and European larch may be planted for fence posts in addition to the timber trees mentioned above.

Protection. — The forest throughout this section is badly in need of protection against stock, fire, wind, etc., but grazing is, perhaps, the most difficult problem to handle, owing to lax custom in the past. Stock laws are more or less uncommon; cattle can roam at will, and it is necessary for owners to fence off their crops instead of confining their cattle.

Grazing should be absolutely prohibited in the woodlands, unless the timber is practically mature and reproduction is not desired. It is impossible to use land advantageously for both grazing and timber production, for the cattle injure the seedlings, pack the soil and prevent reproduction. The trees

FIG. 57. — CLUMP OF CHESTNUT IN MARYLAND.

Before the outbreak of the chestnut bark disease (Endothea parasitica) the chestnut was one of the most highly prized sprout hardwoods. It grows fast, reproduces readily by sprouting and its wood is very durable.

shade out considerable grass so that these two uses are not compatible. If more grass land is required, it would be far better to clear cut some of the poorest forest land to increase the pasturage, rather than to try to serve both purposes on the same piece of ground.

Efficient fire protection is comparatively easy of accomplishment on account of the small size of the average holding and the settled nature of the country. Insect and fungous diseases inflict comparatively little damage, and when present can be controlled by the usual methods.

Utilization. — The question of marketing and utilizing the products of this region, like the management, varies in different parts of the region. Throughout New Jersey, Massachusetts, etc., the markets are unusually stable, and transportation facilities numerous. Here, intensive management and close utilization may be practiced. Thinnings may be made, and the material removed ordinarily finds ready sale as poles, posts, cordwood, etc.

In the Piedmont area, the outlook is not so favorable, on account of the absence of such broad markets. There is no typical forest industry in the region, the larger portion of the lumber being sawed by small mills, many of which are of the portable variety. In the southern Appalachians, however, there are several large hardwood operations cutting largely yellow poplar, oak, hickory, maple, beech, ash, birch and basswood. Arkansas is also an important hardwood center, especially for red gum, oak and hickory.

In the agricultural states like Indiana and Illinois, the possibilities for intensive management and close utilization are excellent. On account of the lack of true forest soil, local forest production will be on a comparatively small scale; intensive forestry will be the rule, and high prices should be obtained. Fence posts and farm timbers find a ready sale, and timber for building and construction purposes can be readily disposed of. Already, plantations established on soils of agricultural richness have shown a net revenue of over five dollars per acre per year. While not as profitable as tilled crops, domestic timber may be produced for the owner's use, with practically no labor outside of getting the plantation started, and a fair profit be obtained beside.

FIG. 58.—YELLOW POPLAR AND WHITE AND RED OAKS.
The central hardwood region is the great source of our hardwood supply.

Special Problems. — A good share of the woodlot holdings, comprising nearly two-fifths of our total remaining forested area in this country, lies in this region, and, consequently, the forestry situation is largely in the hands of the individual. There are no large areas of unseated land to be acquired by the State; there are no areas abandoned by previous owners as in Pennsylvania and Michigan; there are few if any extensive water-sheds within the region which should be covered for the purpose of controlling run-off. It is true, however, that certain districts within the boundaries outlined are subject to overflow at times, but the trouble, for the most part, lies outside this region, and the proper measures must be applied in other districts. The methods of building up the conditions of the run-down woodlot have been described previously in this chapter.

For woodlot production fairly good returns can be assured, provided proper species are used, on account of the character of the soil, its depth, and the precipitation. As far as shelter-belts are concerned, the use of ten to fifteen per cent of the total area in a region subject to hot, drying winds will result in an increase in the agricultural crops, due to the checking of these winds, and to the increased relative humidity supplied by the alternate strips of forest cover.

Future of Forestry in Region. — As has been stated previously, scarcely any region in the United States offers the opportunities for the practice of intensive forestry that the central hardwood region affords. The large cities in the Northeast mean a steady demand for material of all sizes, making close utilization possible; the excellent transportation facilities provide a means of getting the material to the consumer; the comparatively small holdings render intensive management feasible, as soon as the owners become convinced of the financial possibilities of forestry. The Federal Government and various State Forestry organizations are endeavoring to educate the private owners and assist them in every way possible, so that proper forest management, within the central hardwoods region, is a consummation that will doubtless be realized by the coming generation.

BIBLIOGRAPHY.

BAKER, OLIVER. The Forest Problem in a Rich Agricultural County of Ohio. Forestry Quarterly, Volume VI, page 138.

BRYANT, R. C. Railroad Forest Plantation. Forestry Quarterly, Volume V, page 20.

FROTHINGHAM, EARL H. Second Growth Hardwoods in Connecticut. U. S. Forest Service, Bulletin 96.

HAWLEY, R. C. Treatment of Hardwood Lands in Southwestern Connecticut. Forestry Quarterly, Volume V, p. 283.

KELLOGG, R. S. Forest Planting in Illinois. U. S. Forest Service, Circular 81.

RECORD, SAMUEL J. Suggestions to Woodlot Owners in the Ohio Valley Region. U. S. Forest Service, Circular 138.

SPRING, S. N. Forest Planting in Coal Lands in Western Pennsylvania. U. S. Forest Service, Circular 41.

CHAPTER XIX.

PRAIRIE OR FRINGE FOREST.

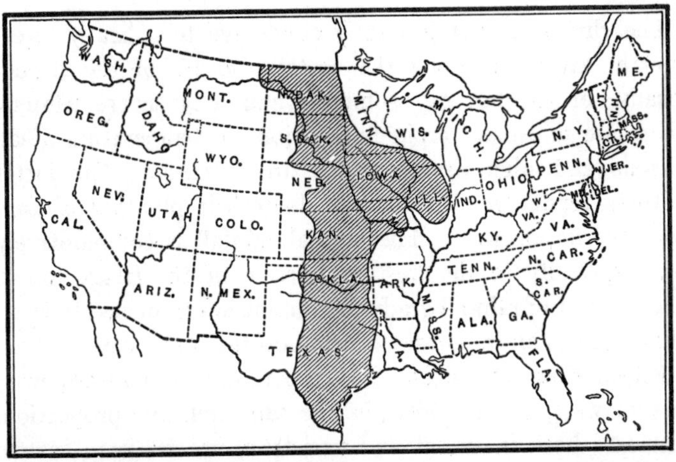

Location and Boundary. — The prairie or fringe forest represents largely the transitional zone from the central hardwood forest to the treeless plains of the prairies. In the extreme north, it is the western continuation of the northern forest and on the south it is the western extension of the southern pine forest. It extends from north to south through the eastern portions of North Dakota, South Dakota, Nebraska, Kansas, Oklahoma and Texas. It includes all of Iowa, southwestern Minnesota, northwestern Illinois and northwestern Missouri. It is very difficult, however, to draw a sharp borderline either on the eastern or western limits of this region.

The topography is uniformly flat or gently rolling, broken by bluffs and gullies along the rivers and stream courses. The forested areas are almost wholly restricted to the bluffs, hillsides and stream banks.

The whole region is essentially an agricultural one; the population is large; and although intensive agriculture is easily the principal occupation, manufacturing is on the rapid increase.

The climate is not generally conducive to a heavy forest growth, excepting along the eastern border of the region, because of the hot dry summers and long severe winters. There are unusually rapid fluctuations in temperature at all seasons and high drying winds are common. The latter features render windbreak and shelterbelt planting of considerable importance. The annual rainfall is comparatively low, especially in the western portion of the fringe forest. The amount of rainfall gradually diminishes from east to west. This fact has an important bearing on the forest cover.

Forest Characteristics. — As the rainfall diminishes from east to west, so the forest, in the same relative proportion, decreases both in extent and variety of species from east to west. Along the eastern fringe there is a considerable portion of the land surface devoted to forest, whereas along the western border of the region, the forest is limited to the stream bottoms and contains very few species.

In all of the portions of the states included within this region, the area devoted to forest is only from about two to fifteen per cent of the total land surface. Due to frequent fires, pasturage and severe cutting for fuel and farm timbers, the native forest has been heavily culled and injured, so that the remaining timber growth is largely scrubby and of an inferior nature.

The character of the forest varies greatly from north to

south but the change consists largely of a substitution of species in the same genera. For example, in the north the forest is characterized by open bur oak growth. In the south there is an increasing amount of southern oaks, osage orange, hackberry, southern elm, etc.

FIG. 59. — PRAIRIE SCENE, HALL CO., NEBRASKA.

The treeless condition of the prairies has been variously ascribed to deficient precipitation, repeated grass fires, too compact soils, tramping of buffaloes, etc., with the question still undecided.

Altogether the principal species of the region are the oaks, among which the most common are the white, bur, red, black and scarlet. There is a great variety of associated species, such as the hickories, walnuts, maples, ash, elm, cottonwood, sycamore, black willow, box elder and basswood. Along the stream courses one finds only the cottonwood, box elder, green ash and willow. The cottonwood extends along the main rivers all the way to the Rocky Mountains. In the central west the oaks and red cedar are the typical trees of the upper bluffs and uplands. Between these two extremes of soil and moisture conditions are found many other species. Even on a bluff 25 to 100 feet in height this variation in species is marked. A few species like the white elm and the ironwood (Ostrya virginiana) are found to the extreme west of the prairies and even to the Black Hills. Others like the cottonwood, aspen and box elder occur as far west as the Rockies.

Since practically the only products of the region are fuelwood, posts and a few other farm timbers, the forest is usually estimated in terms of cords or number of posts. Very little material for saw logs is available except in restricted localities where portable mills can be operated at a profit.

The densely settled region and the development of agriculture render the native and planted timber of the highest importance in the economic life of this section of the country. Most of the land surface has apparently been treeless for a long period of time. This proportion of treelessness increases as one goes west and many attempts have been made to explain this situation.

The following are the most important theories and explanations offered for the treelessness of the prairies: —

1. Insufficient moisture.
2. Constant grazing by the buffaloes and game animals.
3. Soils too compact and heavy.
4. Repeated grass fires.

Most of the evidence seems to support the last-named theory.

Silvicultural Treatment. — A large portion of the present native timber is of sprout origin, especially the oaks, maples, ash, basswood and hickories. The sprout method of regeneration, invigorated from time to time by seedlings, will undoubtedly continue to be the principal means of reproducing the forest. Thinnings for fuel and fence posts can be profitably undertaken from time to time on account of the intensive management possible. Wood products bring attractive prices

FIG. 60. — A WINDBREAK IN THE MIDDLE WEST.

Windbreaks on the north and west sides of fields and homesteads are of great importance on prairies. Cottonwood, willow, catalpa, Norway spruce, white pine, red maple, and white spruce are commonly used.

and make possible very careful and complete silvicultural treatment. The best species to favor in forest management are the oaks, locust, cottonwood, soft maples, ash and hickories.

But the most important phase of silviculture will be devoted to plantations, not only for windbreaks and shelterbelts, but for the commercial growing of fence posts, fuel-wood and general farm timbers. The timber culture act, repealed in 1891 by Congress, was responsible for most of the earlier plantations. Windbreaks were planted on the north and west sides of the homesteads and have proven to furnish efficient protection from the severe winds and snows, not only for the inhabitants but for the cattle and crops. In many cases fertile ground has been planted to catalpa, cottonwood, red maple, white pine and others and annual returns of from $4 to $10 per acre net have been commonly obtained.

In the central portion of the fringe forest the trees recommended for general planting are cottonwood, red and Norway maples, European larch, Norway spruce, white pine, hardy catalpa, honey locust and green ash.

In the northern portion the trees that seem to do best are the green ash, Russian olive, white willow, silver maple, white elm, Norway spruce and European larch.

In the southern portion the best trees for planting are the catalpa, osage orange, green ash, Russian mulberry, honey locust, cottonwood, willows, hackberry, black walnut, red cedar and Kentucky coffee tree.

For windbreak and shelterbelt planting, conifers are preferred because the leaves are absent from the hardwoods during the winter when they are most needed. A triple row of conifers preferably white pine, white cedar, and Norway spruce planted closely together give excellent results.

Protection. — Since most of the holdings are in small woodlots along the bluffs and streams and in plantations, very few phases of protection are of any serious importance. Owing to the lack of continuous forests, the fire problem is reduced to a minimum. A few grass fires occasionally do some damage but they are easily put out. Owing to the intense cultivation of the soil, the formerly large grass fires have been eliminated.

The question of permitting grazing in woodlots and in growing plantations is the most serious phase of forest protection. As in the central hardwood region, particularly in the

Ohio valley, the pasturing of the open timbered areas is very common and has interfered considerably with the proper development and growth of the trees. The forested areas should be devoted exclusively to either pasturage purposes or for the growing of timber. The two are incompatible with each other on the same area for the best results.

Utilization. — Owing to the rapid development of this great agricultural belt, it has been a great consumer of forest products, especially of building lumber and timbers. This is the meeting ground in the consumption of vast quantities of lumber produced in the West, the South and the Lake States.

Forest products bring excellent prices here. Fence posts and fuel-wood will always be in demand. They have, in fact, increased in value to such an extent that the farmers have taken up both the planting of post timbers and the preservative treatment of fence posts. The latter will assume large importance because cheap woods can readily be made to last as long as the cedars, catalpa, locust, white oak and other posts that have largely been brought in from other regions.

Utilization of the raw products of the forest is as complete here as in any region and therefore forest management can be practiced on an intensive basis. Although very few saw logs are produced in these scattered prairie forests, many splendid specimens of walnut logs for furniture, veneers and gunstocks are produced as well as ash, hickory and osage orange stock for vehicle and implement material.

Special Problems. — The two important special problems in forestry in the prairie region will be as follows:

1. Planting of sufficient areas both to raise a supply of fence posts, fuel-wood, etc., to meet the local demand and to furnish efficient shelter to the home and the crops. In furtherance of the fence post supply, wood preservation will be

adopted in direct ratio with the increase in value of this valuable commodity.

2. The elimination of the cattle and sheep from the woodlot. All of the pastured timber holdings are not reproducing satisfactorily and grazing must be eliminated if proper forest conditions are to be established.

Future of Forestry in Region. — Forestry never has nor will reach the importance of agriculture or manufacturing in this region. It is primarily a farming section and practically all of the soils are suited to agricultural development. However, on the steep gullies and bluffs, forests will always have their proper place and should receive adequate care and protection. Forestry is of least importance here of all the regions and is largely a farm woodlot and planting proposition. There is plenty of room for improvement, however, and the good work initiated by many of the state agricultural colleges to promote better care of the forestry interests should be further promoted and encouraged.

BIBLIOGRAPHY.

BAKER, H. P. Native and Planted Timber of Iowa. U. S. Forest Service, Circular 154.

BAKER, H. P., and A. T. ERWIN. Evergreens for the Iowa Planter. Iowa State College, Bulletin 90.

CLOTHIER, G. L. Advice to Forest Planters in Oklahoma, etc. U. S. Forest Service, Bulletin 65.

FETHEROLF, J. M. Forest Planting on the Northern Prairies. U. S. Forest Service, Circular 145.

KELLOGG, R. S. Forest Planting in Western Kansas. U. S. Forest Service, Circular 171.

KELLOGG, R. S. Forest Planting in Illinois. U. S. Forest Service, Circular 81.

MILLER, F. G. Forest Planting in the North Platte and South Platte Valleys. U. S. Forest Service, Circular 109.

MILLER, F. G. Forest Planting in Eastern Nebraska. U. S. Forest Service, Circular 45.

WALDRON, C. B. Windbreaks and Hedges. North Dakota Agricultural College, Bulletin 88.

CHAPTER XX.

NORTHERN ROCKY MOUNTAIN FOREST.

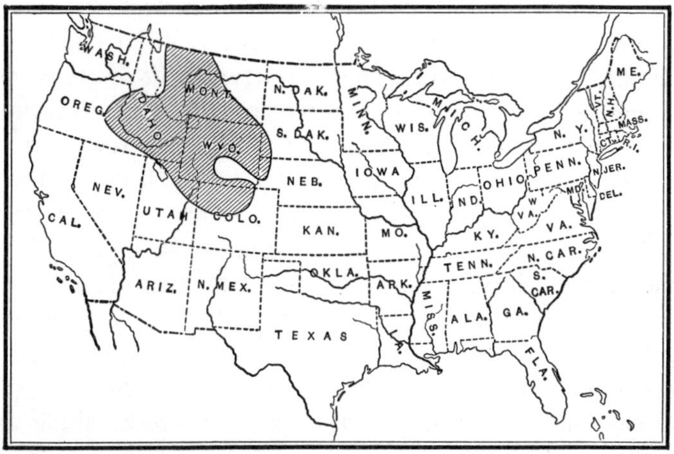

Location and Boundary. — The northern Rocky Mountain forest lies along the northern half of the main continental divide and its outlying ranges and foothills. It includes central Montana, southern Idaho, eastern Oregon, most of Wyoming, northern Colorado, northeastern Utah and small portions in western South Dakota and Nebraska. It is separated from the prairie or fringe forest by a narrow belt of plains which are treeless except along the immediate banks of the rivers and large streams.

This forest region for the most part occupies the mountain slopes above 4000 to 5000 feet in elevation and running up to timber line at 9000 to 11,000 feet above sea level. The lower elevations included in the inter-mountain valleys and plains are treeless, and only require irrigation to make them very fertile agricultural lands.

The topography is uniformly rugged and steep. Most of the principal rivers of the West, flowing both to the Atlantic and Pacific Oceans, have some of their sources in this region. Stream protection to prevent floods and furnish an equal flow for irrigation purposes is therefore of the utmost importance.

The precipitation varies from about 12 to 20 inches per annum. The climate may be characterized as severe, giving a short growing season to the tree growth. High elevations, together with a small amount of rainfall and severe winds, render this region rather unfavorable for the best forest conditions.

A large portion of the forested area has been set aside by the government in National Forests, so that the dual purpose of stream protection and timber production is being well taken care of. Private timber holdings are rather small in extent and have little influence on the national timber supply.

The chief industries of this region are stock raising and mining which are largely developed in connection with the National Forests. In fact many of the National Forests in this region are largely "Grazing Forests."

Forest Characteristics. — The forest is largely composed of yellow pine and lodgepole pine, together with a small amount of Douglas fir, Engelmann spruce and a few minor species, such as alpine fir, limber pine and white bark pine at the high elevations and Colorado blue spruce, aspen, etc.

Probably 90 per cent of the timber is composed of western yellow pine and lodgepole pine, each growing usually in pure stands, the former at the lower elevations and the latter at the higher elevations. In central Montana, Wyoming,

FIG. 61. — TYPICAL STAND OF PURE LODGEPOLE PINE.

This species seldom grows over 2 feet in diameter or 80 feet in height but averages about 8,000 board feet per acre. It is the principal tree of the northern Rocky Mountains.

eastern Idaho and northern Colorado there are vast pure stands of lodgepole pine ranging from 6000 to 9000 feet above sea level. The stands average about 8000 board feet per acre and seldom run over 16,000 feet. This pine never grows over 26 inches in diameter and about 70 to 100 feet in height but the trees stand closely together and acres containing 200 merchantable trees are common. It grows in even-aged stands and reproduces especially well on burns. The cones require considerable heat for opening but when once started by a fire a dense reproduction follows, stands of 50,000 young seedlings per acre being common.

Yellow pine grows in open park-like stands and is usually found along the lower elevations and southerly aspects. Stands of pure yellow pine are characteristic of the Black Hills of South Dakota, southern Idaho and along the eastern slopes of the Rockies in Colorado and Montana. This tree usually occurs in even-aged groups and good merchantable stands run from 5000 to 15,000 feet per acre.

The Douglas fir is occasionally mixed in with both the lodgepole pine and the yellow pine. The occurrence of spruce is governed wholly by the presence of moisture, therefore it is found along streams, seepage flows and even at high elevations where sufficient moisture is afforded. Alpine fir (Abies lasiocarpa) and limber pine (Pinus flexilis) are the typical alpine species found at timber line and just below timber limits.

Altogether the forest in this region may be characterized by unusually slow growth, due to the severe climate, relatively small size and variety of tree growth, abundant natural reproduction of the lodgepole pine, and the broken nature of the forest due to areas above timber line, open parks at high elevations and sage brush valleys and plains between mountain ranges.

Silvicultural Treatment. — Although not of great importance in contributing to the nation's timber supply, the forests of this region are of considerable local importance in supplying mine timbers, railway ties and timbers for local consumption in small town and homestead development.

Fairly intensive methods of silvicultural treatment are therefore justified. Although the lodgepole pine is an intolerant tree growing in even-aged stands and consequently is adapted to management by one of the clear cutting methods, it was found that on account of the windthrow of seed trees left after clear cutting and the lack of demand for the smaller trees, the selection system was best suited to these pure lodgepole pine stands. It was found that by piling and burning the brush on timber sales, and by cutting it down to an elastic diameter limit, satisfactory reproduction is established. On the Deerlodge National Forest in Montana practically every system of management was experimented with and the selection system has recently proven to give the best results. On small timber sales, the government usually allows the logger to cut down to a minimum diameter limit of about ten to twelve inches. The spruce and alpine fir stands are ideally adapted to management under the selection system because they are tolerant trees and grow in all-aged stands. In addition, the forest cover is best maintained for protection purposes under the selection system. Douglas fir in this region is also cut under the selection system. On account of its more rapid growth and the excellent character of its wood, it is favored wherever possible as against lodgepole pine and other associates. Western yellow pine is handled under the selection system. In the Black Hills a rough adaptation of the shelterwood system has been used.

Owing to the large amount and size of the burns in this region, planting must be resorted to in order to fully stock the available areas. Best success is being obtained with western yellow pine, western white pine and Douglas fir. Under ordinary conditions the Forest Service is securing satisfactory natural reproduction on the timber sale cuttings on the National Forests. Practically all forestry work is being done by the government.

Protection. — The whole northern Rocky Mountains have suffered very severely from forest fires in the past. Indians and the railroads have been the most serious causes of fires and on some of the National Forests as much as 60 per cent of the total area has been burned over. Crown fires are most common. When once started, the fire burns everything in its path, due to the dense coniferous stands, steep slopes and hot dry summers. Lodgepole pine has suffered especially in this respect. In the yellow pine types, grass fires are more common and consequently not so disastrous in their effects. The Forest Service is gradually securing most efficient protection through proper brush disposal on logging operations, and by means of look-out points, telephones, trails, and special equipment for preventing and controlling fires.

Windfall is very prevalent in the lodgepole pine forest. Owing to its dense habit of growth and shallow root system it is readily blown over by the wind. In marking trees for felling care is taken not to leave trees alone or improperly protected on exposed sites.

The western pine beetle (Dendroctonus ponderosa) has done enormous damage to the western yellow pine, especially in the Black Hills of South Dakota where large areas have had to be cut to prevent the further spread of this insect. The whole system of management in fact has been moulded around this attack and active measures are being adopted to prevent its spread to other forests and suppress the present outbreak.

Utilization. — As mentioned before, most of the forest products are demanded for local consumption. The mines and railroads require considerable amounts of mine timbers and cross-ties. The Butte mining district requires the output of timber of most of its surrounding forested area. The lodgepole pine, except for its non-durability, makes an excellent mine and tie timber because of its natural size, for these purposes.

The whole region contains large numbers of portable sawmills, cutting lumber and timbers for local consumption. Very little lumber is shipped to outside territory.

Lodgepole pine is used principally for mine props, stulls, lagging, cross-ties and rough lumber. Western yellow pine makes an excellent all around lumber for finishing and general purposes. Douglas fir is used for ties, poles, posts, construction timbers and general lumber.

Stumpage prices are still relatively low but with the increased development of the region, especially in population, the timber resources will be in great demand. The average stumpage prices received at the present time on the National Forests in this region are about as follows: For western yellow pine, $2 to $3; for lodgepole pine, $1.50 to $2.50; for Douglas fir, $2.50; for Engelmann spruce, $3.

Much of the open timbered areas are grazed by cattle and sheep; in fact much of the government revenue is obtained from grazing fees at present from many of the National Forests, especially in Wyoming, central Montana and southern Idaho.

Special Problems. — The forestry problems of this region may be summarized as follows:

1. The regulation of the stream flow for irrigation, reservoir and hydro-electric purposes, as well as to prevent erosion and floods, is of prime importance, especially at the sources of the principal streams flowing out of this section.

The prevention of forest fires, together with the reforestation of the burned areas, are of great immediate importance and are gradually being solved.

Future of Forestry in Region. — Although not one of the most important forest regions in the country, the northern Rockies will always serve an important purpose in supplying timber for the mines, railroads, ranches and the rapidly increasing building in the towns and cities. The forests are already under excellent management by the Forest Service and there is every reason to believe that this region will assume greater importance in fulfilling its part of the forestry program of the country.

Due to the northerly climate, low rainfall, short growing season and high altitudes, the growth of the trees is slow. The rotation for the management of lodgepole pine will be from 80 to 140 years; Douglas fir is somewhat more rapid in its growth and can be handled on a rotation of about 70 to 120 years. It does not grow here nearly as rapidly as on the Pacific slope. Western yellow pine usually does not attain the size here that it does on the Pacific coast or even in the Southwest and will require a rotation of about 80 to 130 years.

BIBLIOGRAPHY.

BATES, C. G. Silvicultural Systems of Management for Central Rocky Mts. Forests. Proc. Soc. Amer. Foresters. Vol. 7, No. 1, 1912.

BATES, C. G., F. B. NOTESTEIN, and PETER KEPLINGER. Climatic Characteristics of Forest Types in the Central Rocky Mts. Proc. Soc. Amer. Foresters. Vol. IX, No. 1, 1914.

BLUMER, J. C. Rocky Mt. Seedling Growth. Forestry Quarterly, No. 2. Vol. IV, 1906.

BROWN, NELSON C. Reproduction of Lodgepole Pine in Relation to its Management. Forestry Quarterly, No. 1. Vol. X, 1912.

CLAPP, E. H. Silvicultural Systems for Western Yellow Pine. Proc. Soc. Amer. Foresters. Vol. VII, No. 2, 1912.

CLEMENTS, F. E. The Life History of Lodgepole Burn Forests. U. S. Forest Service, Bulletin 79.

HODSON, E. R. Silvical Notes on Lodgepole Pine. Proc. Soc. Amer. Foresters. Vol. III, No. 1, 1908.

HODSON, E. R., and J. H. FOSTER. Englemann Spruce in the Rocky Mts. U. S. Forest Service, Circular 170.

LEIBERG, J. B. Forest Conditions in the Absaroka Division of the Yellowstone Reserve. U. S. Geological Survey. Prof. Paper 29.

CHAPTER XXI.

SOUTHERN ROCKY MOUNTAIN FOREST.

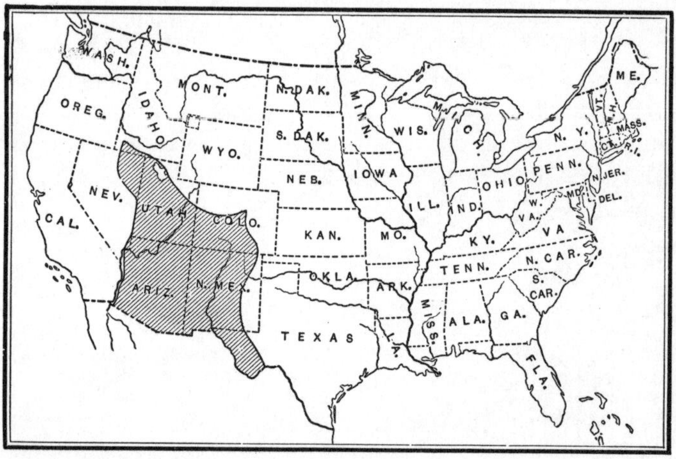

Location and Boundary. — The forest region designated as the Southern Rocky Mountain forest occupies the southern portion of the continental divide, together with its associated high plateaus and ranges.

It is bordered on the south by Mexico and extends north to central Colorado and southern Idaho. It includes the forested portions of Arizona, New Mexico, southern Colorado, most of Utah, northeastern Nevada and a small strip in southern Idaho.

There is no distinct line of demarkation between this region and the northern Rocky Mountain forest but it is separated from it because of the preponderance of the western yellow pine (Pinus ponderosa — also called yellow pine and bull pine) which comprises about 90 per cent of all the merchantable timber of this region. The largest pure pinery in the world is said to exist in New Mexico and Arizona.

The region is characterized by relatively high elevations, broad desert plateaus between the mountain ranges, hot dry climate and rainfall exceedingly low (from 5 to 20 inches per annum).

The principal industry of the region is grazing. Lumbering, however, is on the rapid increase and copper and coal mining are somewhat important in certain centers. Agriculture has been of little consequence in the past, but with several irrigation projects initiated, fruit farming and general ranching are assuming greater importance, especially in Arizona and portions of New Mexico, Utah and southern Colorado.

Most of the forested areas are in the National Forests, as is the case in all the western regions. The government is therefore paying special attention to stream protection and to the permanent upkeep of the sheep and cattle ranges, as well as to timber production.

Forestry is therefore closely allied to the principal industries of the region, including grazing, mining, agriculture and lumbering.

Forest Characteristics. — The commercial stands of timber lie higher in the mountains in this region than in any other region. There is practically no forest growth below elevations of about 5000 eet. At this altitude a few desert plants and trees are first noted. The yellow pine stands occupy the mountain slopes between elevations of about 6000 and 8500 feet. Above this are found the transitional and alpine growths in which many species are present. Timber limits are found higher here than anywhere else in the country, the extreme being an elevation of about 12,400 feet. The forest growth is therefore largely determined by altitudinal and meteorological qualifications.

The forest may be best described by subdividing it into the following distinct types that are typical throughout Arizona and New Mexico especially:

1. Woodland or pinon pine type. This occurs just above the sage brush deserts, usually at elevations of from 5000 to 6000 feet. It is not commercially important and acts as a fringe between the open deserts and the merchantable stands of yellow pine at higher elevations. The principal growth is composed of groups of juniper and pinon pine (Pinus edulis and pinus monophylla) together with some mesquite, (Prosopis juliflora) greasewood (Sarcobatus vermiculatus) and others, among which are Gambel oak, cedar and cypress.

2. Yellow pine type. This occurs usually at elevations of 6000 to 8500 feet and embraces the bulk of the merchantable timber of the region. The stands are characterized by open park-like growth as in the southern pine stands. Occasionally aspen and Engelmann spruce are intermixed at the higher altitudes. Stands of merchantable timber will average from 3000 to 5000 board feet per acre. Yellow pine here averages about 18 inches in diameter and about 70 feet in height. Occasional stands up to 30,000 board feet per acre are found.

3. Transitional or fir type. This is a comparatively narrow belt occurring at elevations of 8500 to 9500 feet and composed largely of a mixture of Douglas fir, white fir, Engelmann spruce, a few yellow pine and limber pine.

4. Alpine type. This is found at the highest limits of timber growth, from about 9,500 up to 12,400 feet in elevation. It is important chiefly for protection purposes. The chief trees are white fir, Arizona fir, bristle-cone pine, limber

pine, aspen and Engelmann spruce. Pure stands of the last-named species are common.

Altogether the southern Rocky Mountain forest is characterized by exceedingly slow growth, dominance of yellow pine, difficulty of securing reproduction and the fact that the principal forest growth is limited to the higher elevations and northerly aspects.

FIG. 62. — OPEN GRAZING LAND AND ALPINE FOREST.
Taken at elevation of 10,000 feet. Holy Cross National Forest, Colorado.

Silvicultural Treatment. — The important question in silvicultural management is the securing of a satisfactory reproduction when the forest is cut over in the government timber sales. The hot dry summers and the comparatively low annual rainfall render reproduction exceedingly difficult.

In handling the yellow pine, the selection system is used.

This tree, however, usually grows in this region, in even-aged groups, so that the group selection system is employed. The tops and brush are lopped and scattered on the ground to assist in shading the soil and therefore in aiding the germination and initial growth of the desired reproduction.

The other types are not of much importance from the standpoint of timber production and therefore receive little silvicultural attention. The pinon type supplies a little fuel and post material and whenever the transitional or fir zone is cut, it is handled on the single selection system, cutting down to a minimum diameter limit of about 10 to 13 inches.

There is considerable over-mature timber in the Southwest and large timber sales are being conducted to remove the over-ripe trees and organize the forests on a definite system of regulating the annual cut according to the annual growth. A rotation of between 160 and 200 years is necessary to produce yellow pine trees of good saw timber size. Measurements show that the average tree of 20 inches in diameter is about 200 years of age.

Protection. — The two important phases of forest protection in this region are stream protection and forest fires. Stream protection has been briefly mentioned already and its importance to irrigation, municipal reservoirs, hydro-electric power plants, as well as to prevent erosion and floods, is at once obvious. Cloudbursts are common in this section, and streams which are dry the greater part of the year are often suddenly turned into raging torrents so that this matter is of especial moment.

Forest fires are usually ground fires burning in the grass on the forest floor. They are rather easily controlled. In the fir and alpine types, however, considerable damage has been done to the forests by crown fires which in places have left thousands of acres a barren waste. In a recent year over 100,000 acres were burned over. Campers and lightning are the most serious causes of fires.

Comparatively little damage is suffered from insects and fungi.

Utilization. — The chief forest product is yellow pine lumber. More than enough is furnished to supply the local demands so that it is shipped to the East and to Southern California. Yellow pine lumber is of excellent quality for all around purposes and is used for a great variety of uses. A great deal of box lumber is cut for the fruit trade.

Altogether utilization is quite complete because the forests are somewhat scattered and broken and the small materials can be successfully utilized for fuel, fence posts, box boards, ranch timbers, mine timbers, etc. Most of the lumber is cut on large logging operations, as opposed to the situation in the northern Rockies.

Experiments have shown that yellow pine may be an important source of our turpentine, rosin and other naval stores in the future when our southern pineries are exhausted.

Special Problems. — The greatest single problem in forestry is to secure a satisfactory natural reproduction of the forests when cut over and to establish new forests on burns by planting and seeding.

The maintenance of proper grazing conditions is also of the highest importance to insure the stable continuance of this large industry.

Stream protection and the control of fires also receives proper attention by the officers of the Forest Service.

Future of Forestry in the Region. — The Southwest will never be an important and great forest region in the sense of being a large producing center and source of the nation's timber supply. But locally forestry will always be important because of its close association with the grazing, mining and agriculture of this region. Several of the National Forests are already on a self-supporting basis and an excellent policy has been inaugurated by the Forest Service for the future care of the forestry problems incident to this region.

BIBLIOGRAPHY.

LEIBERG, J. B., RIXON and DODWELL. Forest Conditions in the San Francisco Mts. Reserve. U. S. Geological Survey. Prof. Paper 22.

MOORE, BARRINGTON. Management of Western Yellow Pine in the Southwest. Forestry Quarterly No. I. Vol. 10, 1912.

PEARSON, G. A. Reproduction of Western Yellow Pine in the Southwest. U. S. Forest Service, Circular 174.

PLUMMER, F. G. and M. G. GOWSELL. Forest Conditions in the Lincoln Forest Reserve, New Mexico. U. S. Geological Survey. Prof. Paper 33.

WOOLSEY, T. S., Jr. Western Yellow Pine in Arizona and New Mexico. U. S. Forest Service, Bulletin 101.

WOOLSEY, T. S., JR. Preliminary Forest Management in the Southwest. Proc. Soc. Amer. Foresters. Vol. IV, No. 1, 1909.

CHAPTER XXII.

PACIFIC COAST FOREST.

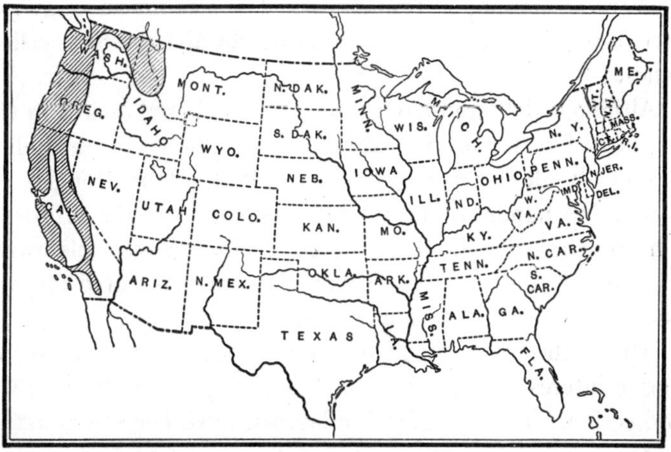

Location and Boundary. — The Pacific coast forest is the most heavily wooded and produces trees of the largest size in the world. It lies chiefly west of the Cascade and Sierra Nevada Mountains in Washington, Oregon and California and also includes northern Idaho and western Montana.

The heavy, dense stands are explained by the favorable climatic influences, such as an equably warm temperature, a heavy precipitation, uniformly distributed, and a long growing season. The precipitation varies from only a few inches in southern California to nearly 100 inches on the Olympic peninsula of the north coast. Elevations run up to over

14,000 feet at several peaks on the main ranges. The soils and slopes are generally very conducive to the best forest growth.

This region contains practically one half of the remaining stand of timber in this country. Approximately 1200 billion board feet of timber are still to be cut in the Pacific coast forests. Oregon contains more of this timber than any other state. Washington is now the leading state in the country in amount of timber cut. Douglas fir, the principal tree, is only exceeded in yearly output by the southern yellow pines.

Although the largest portion of the timbered area is in the National Forests, much of the most valuable stands, including practically all the redwood and the best of the Douglas fir, are in the hands of the private interests and the railroads. The Southern Pacific Railroad Co. is the largest single owner of timber in the country. Its holdings were acquired through the early railroad land grant acts.

All of the states included in this region have taken an active interest in forestry work, especially in the work of fire protection. Even private lumbermen have been very active in this work. Instruction in forestry is also offered in state educational institutions in every state.

Lumbering is by far the most important industry although fruit ranching and the stock industry are of considerable importance as well.

Forest Characteristics. — The forest may be generally characterized as follows:

1. It is almost entirely coniferous.

2. There is a great variety of species, the principal ones being Douglas fir, western yellow pine, redwood, sugar pine, western red cedar, western white pine, hemlock and Sitka spruce.

3. The largest stands in the world are found here. The redwood occurs up to 35 feet in diameter and 350 feet in

FIG. 63. — A HEAVY STAND OF DOUGLAS FIR IN WESTERN WASHINGTON.

Trees are frequently found up to 12 feet in diameter and 250 feet in height. It is adapted to management by the clear-cutting system.

height; the Douglas fir is found up to 12 feet in diameter and 250 feet in height, and the sugar and yellow pines, cedars, spruces and others attain a large size. The average merchantable stand runs from 40,000 to 60,000 board feet per acre in Washington. Many stands are found that run several hundred thousand board feet per acre.

4. Growth is unusually rapid and natural reproduction is readily secured.

There are three distinct subdivisions of the Pacific coast forest, as follows: The northern Douglas fir region, the California sugar pine, yellow pine region and the northern Idaho western white pine region. They are sufficiently important to justify a brief description of each, as follows:

1. Northern Douglas fir region. This is the most important and heavily wooded subdivision. There are several distinct types prevailing in this section which includes for the most part western Washington and western Oregon. On the lower slopes and moist bottomlands is found the best timber, composed largely of Douglas fir. Sitka spruce and western red cedar (giant arborvitæ) occur on the moist situations and swamps, and hemlock and amabilis fir on the better drained soils in association with the Douglas fir. On the upper slopes are found the white, grand and noble firs and hemlock and in some localities, the western white pine, Lawson cypress (Port Orford cedar) Engelmann spruce and others. There are many variations of these types and subtypes in the different parts of this region.

2. The California sugar pine-yellow pine region. This is found largely in California and southern Oregon. It reaches its best development on the western slopes of the Sierras. The climate is warmer, the growing season longer and there is much less rainfall than in western Washington.

There are several types in this sub-region. Along the California coast above San Francisco Bay and along the western slopes of the coast range occur the heavy stands of redwood (Sequoia sempervirens) together with some sugar

FIG. 64. — WESTERN YELLOW PINE IN CALIFORNIA.

This is the most widely distributed conifer in the West. It reaches its best development on the western slopes of the Sierra Mts. It is being cut under the selection system.

pine, incense cedar, Sitka spruce, Douglas fir and grand fir. Most of this is in the hands of the lumbermen.

Throughout the remainder of the region is found a foothill type at elevations up to about 1500 feet. It is of little importance from a commercial viewpoint. The principal species are live oak, digger pine, scrubby yellow pine and much chaparral growing in open poor stands.

Above this type occurs the yellow pine-sugar pine stands embracing the bulk of the commercial timber of California and southern Oregon, outside of the redwood. It runs up to about 9000 feet in elevation. The stands average from about 10,000 to 50,000 board feet to the acre. There are some incense cedar, Douglas fir and white fir intermixed and on the upper slopes and higher elevations there are also white fir, lodgepole pine and red fir. Within this type are 27 isolated bodies of redwood (Sequoia washingtonia). The alpine type is composed of short scrubby specimens of lodgepole, limber and white bark pines and hemlock up to timber line.

3. The northern Idaho-western white pine region. This is found largely in northern Idaho, with extensions in northwestern Montana and northeastern Washington. The principal tree is western white pine with a great variety of others, principal among which are western larch, western red cedar, Douglas fir, yellow pine, Engelmann spruce, etc. The trees do not grow to such a large size as on the Pacific slope, but the mild, moist climate causes a dense heavy growth and stands running up to 150,000 board feet per acre are found. The principal type is a mixed bottomland growth dominated by the white pine (Pinus monticola). Other common types are the larch type, the yellow pine type and the alpine type, containing alpine fir and limber pine.

Silvicultural Treatment. — Very little silviculture has been practiced on the Pacific coast forest for the reason that comparatively few sales have been made on the National Forests. This is so because the private operators are cutting so heavily that up to the present time there has been little demand for the government timber which lies usually in rather inaccessible locations.

FIG. 65. — WESTERN WHITE PINE — WESTERN LARCH TYPE IN NORTHERN IDAHO.

Stands up to 150,000 board feet per acre are common on the moist bottomlands and lower slopes. In the foreground is a squared log, hewed by hand for export.

However, the Douglas fir types are best handled on some clear-cutting system. Seed trees are left singly or in groups and reproduction is readily secured because of the favorable conditions for germination and the thrifty seeding capacity of the Douglas fir. The brush is burned broadcast or in piles. The immense size of the trees and the breakage in felling would not permit of the use of any other system even if it were better adapted to the habits of the tree.

The western yellow pine and sugar pine are best handled under the selection system. The brush is generally piled and burned although lopping and scattering are also resorted to.

The western white pine has been handled in the past by clear cutting and leaving seed trees in groups on high elevations. Owing to its excellent reproductive ability, rapid growth and ability of its seed to germinate on a humous soil, it is also being handled under the selection system.

This whole region is remarkable for the rapid growth and splendid reproductive capacity of its principal trees, thus rendering the practice of forestry a comparatively simple problem. The climatic and soil conditions are highly conducive to forest growth, especially in the northern part of this region. The whole question of forest management seems to depend upon an efficient system of fire protection. That is, if fires are excluded, the forests will reproduce themselves to splendid advantage. In Washington and Oregon, planting will have to be resorted to less than in any other region.

Douglas fir exhibits remarkably rapid growth in second growth stands. In unusual cases, trees have been found to be forty inches in diameter at forty years of age. This tree can be successfully managed on a rotation of from forty to seventy years.

Yellow pine grows to a larger size in California than in any other region. It can be managed along with sugar pine on a rotation of from 60 to 100 years. Second growth redwoods also exhibit a fairly rapid growth. Redwood sprouts to some degree and there is a possibility of using the coppice system to some extent with this tree. Red cedar, hemlock and spruce, the other most important trees of this region, are comparatively slow growers but they are all tolerant trees and can therefore fit in nicely in forest management, often as a second story to Douglas fir which is fairly light demanding in its requirements.

Protection. — As mentioned before fire protection assumes the greatest importance in this region because, with it assured, forest management is a comparatively simple problem. Although characterized by a heavy rainfall, in unusually dry seasons, the fire risk is great, especially in Oregon, Washington and northern Idaho, because of the heavy coniferous stands, dense underbrush and the leaf mould on the forest floor. Crown fires are therefore very disastrous when once started. In California the yellow pine stands are more open in their growth so that grass fires are the more common kind.

The lumbermen of this region have formed several protective associations and are very much alive to the installation of proper protective measures. They have coöperated with the Forest Service, and the region is now being handled most efficiently by means of fire patrol, lookout stations and towers, fire trails and telephones. All of the states have good fire laws on their statutes and altogether this region is better administered from the standpoint of fire protection than any other region.

Trametes pini and other fungi have done a lot of damage to the Douglas fir, western yellow pine and other trees. The black rot attacks the hemlock very seriously. Owing to the moist conditions here, the work of the fungi is more serious than in any other region.

Insects have done considerable damage in California but the Forest Service is combatting them with success.

The regulation of the stream flow for irrigation, reservoir and electric power plants is of prime importance here, owing to the population and the number of relatively large towns and cities. Irrigation for the fruit districts of California, Washington and Oregon is of especial interest.

Utilization. — This region is really the great " storehouse " of the nation's timber supply, owing to the fact that nearly one half of the remaining timber supply of the country is to be found here. Up to the present time very rough and extensive utilization has characterized the lumber operations in this section. Lumber has been cheap and plentiful, with the natural result that most trees are cut down to a diameter of only 12 inches or more in the tops and only the best material is utilized.

Some of the largest logging operations in the country are located here. Steam logging is typical throughout the forests because horses are not able to move the enormously large logs. The logging operations are characterized by wastefulness in high stumps, large and broken tops and the damage to the young growth.

The principal lumber trees at present are Douglas fir, western yellow pine, redwood, sugar pine, spruce, red cedar, hemlock, incense cedar, yellow cypress, fir and white pine. Owing to their relative abundance, lumber of all species is cheap and therefore close utilization is impossible. However, several paper plants, wood distillation plants and boxboard factories are being installed in connection with the larger plants to utilize the waste. The chief forms of waste are the large slabs, sawdust and the tops and broken material in the woods.

Douglas fir is largely cut into lumber, construction timbers and railroad ties. Its wood is strong, durable and of high technical value. Cedar is the great shingle wood, producing about 75 per cent of all the shingles used in this country. Western yellow pine and sugar pine produce high-grade lumber. Redwood lumber is the most fire resistant of any wood and therefore furnishes excellent house construction lumber. It is also used for grape stakes, boxboards and railroad ties.

To summarize, the situation in utilization is characterized as follows:

(1) This is a great lumber production center.

(2) It produces a large number of high-grade timber trees.

(3) It produces the largest size construction timbers available anywhere.

(4) Utilization is not complete because of the great supply of lumber and therefore its cheapness.

Special Problems. — The special problems are as follows:

(1) The establishment of an efficient fire protection. The high value of the large timbered areas renders the expenditure of a few cents per acre every year a cheap form of fire insurance.

(2) The closer utilization of forest products. This will be solved with the increased value of lumber and the introduction of means of more completely utilizing the wood for paper, chemical distillates and products, boxboards, fruit packages, veneers, cooperage and the manufacture of small wooden articles.

Future of Forestry in Region. — The future of this region is unusually good because of its abundant timber supply, rapid growth of its principal species, rapidly growing population, large percentage of true forest soils and a general public interest in the welfare of lumbering, the chief industry which depends upon the permanence of the forest for its existence. The manifestation of this interest through the passing of good forest and fire laws, the establishment of forestry courses in the state educational institutions and the formation of the fire protective associations, together with the work of the Forest Service, all assure a most hopeful future for the most important single forest region in the country.

BIBLIOGRAPHY.

Berry, Swift. Notes on Management of Redwood Lands. Vol. VI, No. 1. Proc. Soc. Amer. Foresters, 1911.

Burns, Findley. The Olympic National Forest; Its Resources and Their Management. U. S. Forest Service, Bulletin 89.

Burns, Findley. The Crater National Forest; Its Resources and Their Conservation. U. S. Forest Service, Bulletin 100.

Cooper, A. W. Sugar Pine and Western Yellow Pine in California. U. S. Forest Service, Bulletin 69.

Frothingham, E. H. Douglas Fir; A Study of the Pacific Coast and Rocky Mt. Forms. U. S. Forest Service, Circular 150.

Greeley, W. B. A Rough System of Management for Reserve Lands in the Western Sierras. Proc. Soc. Amer. Foresters. Vol. II, No. 1, 1907.

Kirkland, B. P. Working Plans for National Forests in the Pacific Northwest. Proc. Soc. Amer. Foresters. Vol. VI, No. 1, 1911.

Leiberg, J. B. Forest Conditions in the Northern Sierra Nevada, California. U. S. Geological Survey. Prof. Paper 8.

Mason, D. T. Management of Western White Pine. Proc. Soc. Amer. Foresters. Vol. IX, No. 1, 1914.

Munger, T. T. The Growth and Management of Douglas Fir in the Pacific Northwest. U. S. Forest Service, Circular 175.

Munger, T. T. Natural vs. Artificial Regeneration in the Douglas Fir Region of the Pacific Coast. Proc. Soc. Amer. Foresters. Vol. VII, No. 2, 1912.

Sudworth, G. B. Forest Trees of the Pacific Slope. U. S. Forest Service, Unnumbered Bulletin.

APPENDIX.

ORIGINAL AND PRESENT FOREST AREAS IN THE UNITED STATES.

Region.	Original forest.		Present forest.			
	Area.	Stand.	Area.	Stand.	Per cent of original area.	Per cent of original stand.
	Million acres.	Billion ft. B. M.	Million acres.	Billion ft. B. M.	Per cent.	Per cent.
Northern	150	1000	90	300	60	30
Southern	220	1000	150	500	68	50
Central	280	1400	130	300	46	21
Rocky Mt.	110	400	100	300	91	75
Pacific Slope	90	1400	80	1100	89	79
	850	5200	550	2500	65	48

Forest Service Circular 166. The Timber Supply of the United States by R. S. Kellogg.

Uses of the Principal American Species.

Showing also the common and scientific names, distribution and maximum sizes.

Common and scientific names.	Maximum diameter and height.*	Distribution.	Qualities and uses.
		Conifers.	
White or Weymouth pine (*Pinus strobus*).	6×250	Maine to Minnesota and along Appalachians to Georgia.	Light, soft and easily worked. Most useful American wood. Not specialized in its uses.
Longleaf, Georgia or hard pine (*Pinus palustris*).	3×120	Virginia to Texas along coastal plain.	Hard, heavy, strong, durable, and resinous. Used in construction, ties, flooring, and general lumber. Source of our naval stores.
Shortleaf or yellow pine (*Pinus echinata*₂.)	4×120	New York south to Texas. Also in Mississippi Valley up to Missouri and Illinois.	Fairly hard, heavy, and strong. Used with longleaf pine for the same purposes.
Cuban or slash pine (*Pinus heterophylla*).	3×115	South Carolina to Louisiana along the coast.	Same qualities and uses as longleaf pine.
Loblolly or old field pine (*Pinus taeda*).	5×150	From New Jersey along coast to Texas and Arkansas.	Fairly heavy, coarse-grained, fairly durable. Used for general lumber and box boards.
Western yellow or bull pine (*Pinus ponderosa*).	7×200	Found in every western mountain state.	Fairly heavy, close-grained, easily worked. Used for lumber, construction timbers, ties and mine timbers.
Red or Norway pine (*Pinus resinosa*).	3×120	Maine to Minnesota and south to Pennsylvania.	Medium between white and yellow pines in qualities. Used for lumber.
Sugar pine (*Pinus lambertiana*).	12×250	South Oregon and along Sierras in California.	Same qualities and uses as white pine.
Western white pine (*Pinus monticola*).	8×220	British Columbia to California. Chiefly in North Idaho and western Montana.	Same qualities and uses as white pine.
Lodgepole pine (*Pinus murrayana*).	2½×100	Alaska to California and Colorado.	Light, soft, weak, brittle, not durable wood. Used locally for ties, mine timbers, and general lumber.
Red spruce (*Picea rubens*).	3×100	From valley of St. Lawrence south along Appalachians to North Carolina.	Light, soft, close-grained, not durable. Used mostly for paper pulp, sounding boards and dimension timbers.
Sitka spruce (*Picea silchensis*).	16×200	Alaska to North California on Pacific coast.	Light, soft, close and straight grained. Used for lumber, cooperage, boats, pulp, and woodenware.
Engelmann spruce (*Picea engelmanni*).	5×150	Through Rocky Mountains from Arizona to British Columbia.	Light, soft wood. Used for general lumber for local purposes.
Douglas fir (*Pseudotsuga taxifolia*).	12×250	Found in all western mountain states.	Heavy, hard, durable, strong wood. Used for lumber, construction, ties, shipbuilding.
Hemlock (*Tsuga canadensis*).	4×125	Maine to Minnesota and south on Appalachians to Georgia.	Soft, weak, brittle wood. Used for coarse lumber and small dimension timbers.

* Measurements are given in feet unless otherwise noted.

USES OF THE PRINCIPAL AMERICAN SPECIES. (*Continued*)

Common and scientific names.	Maximum diameter and height.	Distribution.	Qualities and uses.
		Conifers.	
Western hemlock (*Tsuga heterophylla*).	8×250	Alaska to California and Montana.	Light, hard, tough, not durable wood. Used for rough lumber and construction timbers.
Tamarack or larch (*Larix laricina*).	20 in. × 60 ft.	Newfoundland to Minnesota, south to Pennsylvania.	Hard, heavy, strong and durable. Used for ties, posts, poles, ships, and rough lumber.
Western larch (*Larix occidentalis*).	8×250	British Columbia to Oregon and Montana.	Very hard, heavy, strong, durable and close-grained. Used for ties, construction timbers, and lumber.
Balsam fir (*Abies balsamea*).	30 in. × 80 ft.	Newfoundland to Minnesota, south to Virginia.	Light, soft, weak, perishable, coarse-grained. Used for pulp, boxes and generally sold as spruce.
Amabilis or white fir (*Abies amabilis*).	6×250	On Pacific Coast. Oregon to British Columbia.	Light, rather soft and weak. Used for rough lumber, packing cases, etc.
Noble fir or larch (*Abies nobilis*).	8×250	On Pacific Coast, Washington to California.	Light, hard, strong wood. Used for lumber, construction and cases.
Red fir (*Abies magnifica*.)	10×200	Western slopes of the Sierras.	Light, soft, rather weak. Used for rough lumber, construction, and cases.
Bald cypress (*Taxodium distichum*).	12×150	Delaware to Texas along coast and up to Illinois and Indiana.	Light, soft, durable, very workable. Used for ties, posts, cooperage, doors, shingles and inside trim.
Big tree or redwood (*Sequoia washingtoniana*).	35×320	Western slopes of the Sierras in California.	Light, soft, durable, weak wood. Used for shingles, grape stakes, ties, and general lumber.
Redwood (*Sequoia sempervirens*).	20×350	Northern California coast region.	Same qualities and uses as the big-tree.
Western red cedar or giant arbor-vitæ (*Thuja plicata*).	15×200	Alaska to California and Montana.	Light, soft, very durable and brittle. Used for shingles, posts, poles, cooperage and lumber.
Arbor-vitæ or white cedar. (*Thuja occidentalis*).	18 in. × 60 ft.	Nova Scotia to Minnesota, south to North Carolina.	Light, soft, brittle, very durable. Used for shingles, poles, posts.
Incense cedar or white cedar (*Libocedrus decurrens*).	8×200	Southern Oregon and California.	Soft, light, weak, brittle, but very durable. Used for shingles, and general lumber.
Port Orford cedar or Lawson's cypress (*Chamaecyparis lawsoniana*).	12×200	Along Pacific Coast, Oregon and California.	Light, hard, strong, and durable. Used for flooring, ties, ships, matches and lumber.
Red cedar (*Juniperus virginiana*).	4×100	Found everywhere east of the Rocky Mountains.	Light, soft, and close-grained. Used for pencils, cabinets, posts and chests.

USES OF THE PRINCIPAL AMERICAN SPECIES. (*Continued.*)

Common and scientific names.	Maximum diameter and height.	Distribution.	Qualities and uses.
		Hardwoods.	
White oak (*Quercus alba*).	4×100	Maine to Minnesota, south to the Gulf.	Hard, strong, durable, and seasons well. Used for ships, furniture, ties, cooperage, veneers, flooring and cabinet work.
Other oaks classed as white oaks and used for the same purposes are post, bur, rock, swamp white, cow and live oaks. All of these are found in the East.			
Red oak (*Quercus rubra*).	4×150	Nova Scotia to Minnesota and south to Georgia and Kansas.	Similar to white oak, but not as hard, strong or as durable. Used for furniture, ties, interior finish and general hardwood lumber.
Other oaks classed as red oaks and used for the same purposes are scarlet, pin, black, Spanish and water oaks. All are found in the East.			
Sugar or hard maple (*Acer saccharum*).	5×125	Found everywhere east of the prairies.	Hard, heavy, strong, tough, but not durable. Used for furniture, cabinets, tools, implements, instruments and flooring.
White ash (*Fraxinus americana*).	5×100	Newfoundland to Minnesota, south to Gulf.	Heavy, hard, tough, seasons well. Used for all kinds of implements, furniture, ball-bats, handles, vehicles, etc.
Shagbark hickory (*Hicoria ovata*).	4×125	Maine to Minnesota, south to Gulf.	Very heavy, hard, tough, and strong. Used for axe and tool handles, implements, vehicles, etc.
Three other hickories (*Hicoria alba, glabra,* and *minima*) are also commonly used as hickory and, in general, exhibit the same qualities as the shagbark. All grow in the East.			
Yellow or red birch (*Betula lutea*).	4×100	Newfoundland to Minnesota, south to North Carolina.	Heavy, very strong and hard and close-grained. Used for furniture, hubs, handles, flooring, veneers and interior finish.
Yellow poplar or tulip (*Liriodendron tulipifera*).	10×200	Vermont to Florida, west to Michigan and Arkansas.	Light, soft, even texture, seasons well, not very durable. Used for interior finish, boats, woodenware and general hardwood lumber.
Red gum or sweet gum (*Liquidambar styraciflua*).	5½×150	Connecticut to Florida, west to Missouri and Texas.	Fairly heavy, satiny, difficult to season, cross-grained. Used mostly for veneers, cooperage, furniture, interior finish.

USES OF THE PRINCIPAL AMERICAN SPECIES. (*Concluded.*)

Common and scientific names.	Maximum diameter and height.	Distribution	Qualities and uses.
		Hardwoods.	
Black walnut (*Juglans nigra*).	6×150	New Brunswick to Minnesota and south to the Gulf.	Light, soft, even-grained, seasons well, yields a beautiful polish. Used for furniture, veneers, cabinets, gun-stocks, and fancy hardwood articles.
Butternut or white walnut (*Juglans cinerea*).	4×125	New Brunswick to Minnesota and south to Georgia and Arkansas	Light, soft, coarse-grained. Used as substitute for black walnut.
Chestnut (*Castanea dentata*).	12×100	Maine to Michigan, south to Mississippi and Georgia.	Light, soft, coarse-grained, durable. Used for ties, poles, posts, mine props and general lumber. Also for tannin.
Beech (*Fagus atropunicea*).	4½×120	Nova Scotia to Wisconsin south to Florida and Texas.	Very hard, heavy, strong and tough, not durable, difficult to season. Used for chairs, handles, woodenware, cooperage, flooring, shoe lasts, etc.
White elm (*Ulmus americana*).	11×120	Newfoundland to Rocky Mountains, south to the Gulf.	Very heavy, hard, tough, cross-grained. Used for implements, hubs, wagon parts, cooperage, handles, etc.
Basswood or linden (*Tilia americana*).	4½×140	New Brunswick to Minnesota, south to Texas and Georgia.	Light, soft, seasons excellently, even-grained, tough. Used for woodenware, excelsior, cooperage, veneer backing, trunks and general lumber.
Black or yellow locust.	3½×90	Pennsylvania to Georgia, west to Minnesota and Oklahoma.	Very hard, heavy, strong and durable. Used for ships, insulator pins, wagon stock, posts and certain specialized uses.
Sycamore or plane (*Platanus occidentalis*).	12×170	Found in every state east of the central prairies.	Heavy, hard, not durable, cross-grained. Used for furniture, butcher's blocks and small wooden articles and woodenware.
Black or red cherry (*Prunus serotina*).	5×110	Same distribution as sycamore.	Strong, hard, close-grained, satiny, very durable. Excellent cabinet wood. Used for fine furniture, interior finish, instruments, cases, clocks, etc.
Cottonwood (*Populus deltoides*).	8×100	Found in every state east of the Rocky Mountains.	Soft, light, weak, cross-grained. Used for crates and boxes, cooperage and cheap lumber.
Swamp cottonwood or cottonwood (*Populus heterophylla*).	3×130	Connecticut to Georgia, west to Arkansas. Found mostly in South.	Soft, light, even, straight-grained. Seasons well. Used for general lumber purposes, boxes and crating, veneers and cooperage.

LOG RULES

Table I. — Scribner Log Rule.

(Decimal " C.")

Showing contents in board feet for the various log lengths.

Diameter.	8	10	12	14	16
Inches.	Bd. ft.	Bd. ft.	Bd. ft.	Bd. ft.	Bd. ft.
6	0.5	1	1	1	2
7	1	1	2	2	3
8	1	2	2	2	3
9	2	3	3	3	4
10	3	3	3	4	6
11	3	4	4	5	7
12	4	5	6	7	8
13	5	6	7	8	10
14	6	7	9	10	11
15	7	9	11	12	14
16	8	10	12	14	16
17	9	12	14	16	18
18	11	13	16	19	21
19	12	15	18	21	24
20	14	17	21	24	28
21	15	19	23	27	30
22	17	21	25	29	33
23	19	23	28	33	38
24	21	25	30	35	40
25	23	29	34	40	46
26	25	31	37	44	50
27	27	34	41	48	55
28	29	36	44	51	58
29	31	38	46	53	61
30	33	41	49	57	66
31	36	44	53	62	71
32	37	46	55	64	74
33	39	49	59	69	78
34	40	50	60	70	80
35	44	55	66	77	88

From Forest Service Bulletin 36, Woodsman's Handbook.

The total scale is obtained by multiplying the figures in this table by 10. Thus, the contents of a 6-inch 8-foot log are given as 0.5, so the total scale is 5 board feet. A 30-inch 16-foot log is given as 66, or a total scale of 660 board feet.

TABLE II. — DOYLE LOG RULE.

Showing contents in board feet for the various log lengths.

Diameter in inches.	Length in feet.				
	8	10	12	14	16
	Contents in board feet.				
6	2.0	2.5	3.0	3.5	4
7	4.5	5.6	6.8	7.9	9
8	8	10	12	14	16
9	12	16	19	22	25
10	18	23	27	32	36
11	24	31	37	43	49
12	32	40	48	56	64
13	40	50	61	71	81
14	50	62	75	88	100
15	60	75	91	106	121
16	72	90	108	126	144
17	84	106	127	148	169
18	98	122	147	171	196
19	112	141	169	197	225
20	128	160	192	224	256
21	144	181	217	253	289
22	162	202	243	283	324
23	180	226	271	313	359
24	200	250	300	350	400
25	220	276	331	386	441
26	242	302	363	423	484
27	264	330	397	463	530
28	288	360	432	504	576
29	312	391	469	547	625
30	338	422	507	591	676
31	364	456	547	638	729
32	392	490	588	686	784
33	420	526	631	736	841
34	450	562	675	787	900
35	480	601	721	841	961
36	512	640	768	896	1024

TABLE III. — THE CHAMPLAIN LOG RULE.

Showing contents in board feet for the various log lengths.

Diameter, inches.	8	10	12	14	16
	Contents in board feet.				
4	4	5	6	7	8
5	7	9	10	12	14
6	11	14	17	19	22
7	16	20	24	28	32
8	21	27	32	38	43
9	28	35	42	49	56
10	35	44	53	62	70
11	43	54	65	76	87
12	52	65	78	92	105
13	62	78	93	109	124
14	73	91	109	127	146
15	84	105	126	147	168
16	97	121	145	169	193
17	110	137	165	192	219
18	124	155	186	217	247
19	139	173	208	242	277
20	154	193	231	270	308
21	171	213	256	299	341
22	188	235	282	329	376
23	206	258	309	361	412
24	225	282	338	394	450
25	245	306	368	429	490
26	266	332	399	465	532
27	287	359	431	503	575
28	310	387	465	542	620
29	333	416	499	583	666
30	357	446	535	625	714
31	382	477	573	668	764
32	407	509	611	712	814
33	434	543	651	760	868
34	462	577	692	808	923
35	490	612	735	857	980

From Bulletin 102, Vermont Agricultural Experiment Station. By Professor A. L. Daniels.

Sample Tally Sheet Used in Timber Estimating.

Locality

Type *Date* ________ *No.*

Diameter, breast-high.	White pine.	Red oak.	Hemlock.	Beech.	Ash.						Totals.
Inches. 10											
11											
12											
13											
14											
15											
16											
17											
18											
19											
20											
21											
22											
23											
24											
25											
26											
27											
28											
29											
30											
31											
32											

Sample Sheet Used in Recording Tree Measurements in Stem Analysis to Determine the Rate of Growth of Trees.

NEW YORK STATE COLLEGE OF FORESTRY.

Locality *Type* *Species*

Date *Tree Class* *D. B. H.* *No.*

Cross sec.	Distance in inches on average radius from center to ring. (Count in, measure out.)																			
	10	20	30	40	50	60	70	80	90	100	110	120	130	140	150	160	170	180	190	200
St.																				
2																				
3																				
4																				
5																				
6																				
7																				
8																				
9																				
10																				

Cross sec.	Distance in inches on average radius from center to ring.										Age, years.	Length, sec. ft.	D. i. B. in.	D. o. B. in.	Sap, in.	Sap, rings.
	210	220	230	240	250	260	270	280	290	300						
St.																
2																
3																
4																
5																
6																
7																
8																
9																
10																

Tot. hgt. ft.,
Cl. lgth. ft.,
M. log. ft.,
Crn. ft.,
Form factor,
V. cu. ft.,
V. wd. cu. ft.,
M. v. cu. ft.,
M. v. stan.,
M. v. cds.,
M. v. bd. ft.,
Per cent mer.,
Per cent bark,
Per cent sap,

TABLE IV.—VOLUME TABLE, IN BOARD FEET, FOR WHITE PINE IN MASSACHUSETTS.

Scaled from rules made by mill tallies. Volume to 4-inch top and ½-foot stump.

Diameter, breast-high (inches).	Total height (feet).							
	30	40	50	60	70	80	90	100
	Bd. ft.	Bd. ft.	Bd. ft.	Bd. ft.	Bd. ft.	Bd. ft.	Bd. ft.	Bd. ft.
5	10	....	....	....	....	......	....	....
6	15	20	30	....	....	......	....	....
7	20	30	40	50	65	......	....	....
8	25	35	50	65	85	......	....	....
9	30	45	60	80	105	115	....	....
10	40	55	75	95	125	145	....	....
11	....	65	90	115	145	170	200	230
12	....	75	105	135	165	200	230	260
13	....	85	120	155	190	235	260	295
14	....	100	140	175	215	265	300	335
15	....	115	160	200	245	300	340	375
16	....	....	180	230	275	335	380	420
17	....	....	....	260	310	370	425	470
18	....	....	....	295	350	410	475	530
19	....	....	....	335	390	455	530	600
20	....	....	....	380	435	505	580	660
21	....	....	....	....	480	550	635	720
22	....	....	....	....	520	595	680	780
23	....	....	....	....	565	640	730	835
24	....	....	....	....	600	690	780	890
25	....	....	....	....	645	740	830	940
26	....	....	....	....	....	......	885	995
27	....	....	....	....	....	......	940	....

From "The White Pine in Massachusetts." By permission of the Mass. State Forester.

TABLE V.—YIELD TABLE FOR WHITE PINE.

Age (years).	Quality I.			Quality II.			Quality III.		
	1-inch boards.	Cords.	Cubic feet.	1-inch boards.	Cords.	Cubic feet.	1-inch boards.	Cords.	Cubic feet.
25	10,825	25.1	2080	6,750	16.4	1300	3,975	10.8	750
30	19,900	44.0	3750	12,500	31.2	2740	7,500	18.2	1400
35	31,150	60.4	5420	24,400	49.0	4375	16,950	35.8	3035
40	40,650	70.6	6590	32,800	58.0	5300	25,200	46.2	4080
45	49,350	78.0	7420	40,600	64.8	6075	32,100	51.8	4785
50	55,150	84.2	8035	46,500	70.0	6725	37,550	56.6	5475
55	59,650	89.2	8575	50,550	74.8	7200	42,100	60.8	6015
60	63,600	93.4	9075	53,200	79.2	7655	44,550	64.6	6340
65	67,050	97.2	9550	56,600	83.0	8050	46,150	68.4	6550

From "The White Pine in Massachusetts." By permission of the Mass. State Forester.

TABLE VI. — MONEY YIELD TABLE FOR WHITE PINE.

Manufactured and stumpage values.

Age (Years).	Quality I.					Quality II.					Quality III.				
	Volume.	Per M ft.	F. O. B. value.	Per M ft.	Stumpage.	Volume.	Per M ft.	F. O. B. value.	Per M ft.	Stumpage.	Volume.	Per M ft.	F. O. B. value.	Per M ft.	Stumpage.
	Bd. ft.					Bd. ft.					Bd. ft.				
25	10,825		$173.20	$6.00	$65.00	6,750		$108.00	$6.00	$40.50	3,975		$63.60		$23.85
30	19,900	$16.00	318.49		119.40	12,500	$16.00	200.00		75.00	7,500	$16.00	120.00	$6.00	45.00
35	31,150		498.40		249.20	24,400		439.20		195.20	16,950		271.20		101.70
40	40,650		791.70	$8.00	325.20	32,800		590.40	$8.00	262.40	25,200		403.20		201.60
45	49,350	$18.00	888.30		394.80	40,600	$18.00	730.80		324.80	32,100		577.80	$8.00	256.80
50	55,150		992.70	$10.00	551.50	46,500		837.00		465.00	37,550		676.00		300.40
55	59,650		1193.00		596.50	50,550		910.00	$10.00	505.50	42,100	$18.00	757.80		336.80
60	63,600	$20.00	1272.00	$12.00	763.20	53,200	$20.00	1064.00		532.00	44,550		802.00	$10.00	445.50
65	67,050		1341.00		804.50	56,600		1132.00		566.00	46,150		830.70		461.50

From "The White Pine in Massachusetts." By permission of the Mass. State Forester.

TABLE VII. — APPROXIMATE AVERAGE RATE OF GROWTH FOR NORTHERN FORESTS.

Species.		Locality.	Forest type.	Age.											
				30 years.		50 years.		80 years.		100 years.		150 years.		200 years.	
Common name.	Botanical name.			Diameter.	Height.	Diameter.	Height.	Diameter.	Height.	Diameter.	Height.	Diameter.	Height.	Diameter.	Height.
				In.	Ft.	In.	Ft.	In.	Ft.	In.	Ft.	In.	Ft.	In.	Ft.
Arborvitæ.	Thuja occidentalis.	Michigan.	Swamp.	2.2	15	3.8	28	5.7	34	6.9	38	9.8	45	12.3	50
Aspen.	Populus tremuloides.	Maine.	Mixed hardwood.	6.0	45	9.4	65	12.0	76	....	...	....	...	....	...
Basswood.	Tilia americana.	Michigan.	Mixed hardwood.	4.5	42	8.1	51	13.2	67	15.7	72	19.1	75	....	...
Beech.*	Fagus atropunicea.	Michigan.	Mixed hardwood.	1.8	19	3.8	35	8.0	68	10.2	73	13.6	74	16.6	75
Birch, paper.	Betula papyrifera.	Maine.	Paper birch — Seedling.	5.3	44	8.0	62	10.2	78	....	...	....	...	....	...
		New Hampshire.	Paper birch — Sprout.	5.6	49	7.9	70	....	...	....	...	....	...	....	...
Birch, yellow.	Betula lutea.	New York.	Mixed hardwood.	2.6	24	4.5	35	7.4	52	9.4	59	14.5	69	19.4	75
Elm, white.	Ulmus americana.	No data.	....	....	...	....	...	....	...	....	...	....	...	....	...
Fir, balsam.	Abies balsamea.	New York.	Dry swamp.	2.3	20	5.2	42	7.2	53	7.6	54	8.3	57	....	...
Hemlock.*	Tsuga canadensis.	Michigan.	Hemlock — hardwood.	1.3	9	2.9	20	5.7	37	7.8	45	13.4	63	18.4	73
Maple, sugar.*	Acer saccharum.	Michigan.	Mixed hardwood.	1.9	21	3.8	35	7.0	50	9.0	64	14.0	72	....	...
Pine, jack.	Pinus divaricata.	Minnesota.	Jack pine — Quality I.	6.2	40	8.5	50	....	...	....	...	....	...	....	...
Pine, red.	Pinus resinosa.	Wisconsin.	Ravine — Quality I.	5.8	45	10.3	75	15.3	90	17.6	95	22.6	103	26.1	108
Pine, white.	Pinus strobus.	New York.	Lower slopes.	5.8	42	10.6	60	16.6	82	20.0	92	26.0	106	29.8	115
		Minnesota.	White and red pine.	5.3	48	9.2	71	13.0	82	14.8	85	18.3	91	21.7	96
Spruce, black.	Picea mariana.	Minnesota.	Swamp.	....	...	1.1	9	2.6	17	3.0	20	4.1	26	4.7	29
Spruce, red.	Picea rubra.	Maine.	Spruce slope.	2.4	21	5.3	36	7.2	44	....	...	....	...	....	...
Spruce, white.	Picea canadensis.	Minnesota.	Swamp.	....	...	2.3	22	4.9	41	6.0	46	9.0	60	....	...
Tamarack.	Larix laricina.	Minnesota.	Swamp.	2.9	32	5.5	50	6.9	62	7.8	66	11.0	75	....	...

* These species were undoubtedly suppressed for some years. *Note.* — The diameter given is 4.5 feet from the ground.

From Report of the National Conservation Commission Rate of Forest Growth, by E. A. Zeigler, page 205.

TABLE VIII. — APPROXIMATE AVERAGE RATE OF GROWTH FOR CENTRAL HARDWOOD FORESTS.

Species.		Locality.	Forest type.	Age.											
				30 years.		50 years.		80 years.		100 years.		150 years.		200 years.	
Common name.	Botanical name.			Diameter.	Height.	Diameter.	Height.	Diameter.	Height.	Diameter.	Height.	Diameter.	Height.	Diameter.	Height.
				In.	Ft.	In.	Ft.	In.	Ft.	In.	Ft.	In.	Ft.	In.	Ft.
Catalpa.*	Catalpa speciosa.	Illinois.	Farm plantations.	9.0	45	...	...	...	...	...	...	...	...	...	...
Chestnut.	Castanea dentata.	Maryland.	Hardwood, seedling.	6.0	33	11.2	64	17.2	84	20.1	91	...	...	...	...
			Hardwood, sprout.	9.3	57	13.4	77	18.0	90	19.8	93	...	...	...	...
Mockernut (hickory).	Hicoria alba.	Mississippi.	Mixed hardwood.	3.3	...	6.0	...	10.5	...	13.3	...	20.0	...	26.0	...
Larch, European.*	Larix europea.	Illinois.	Farm plantations.	8.8	45	...	...	...	...	...	...	...	...	...	...
Locust, black.	Robinia pseudacacia.	Kentucky.	Close to open stands.	7.1	36	11.5	44	...	...	...	...	...	...	...	...
Maple, silver.*	Acer saccharinum.	Illinois.	Farm plantations.	9.0	55	...	...	...	...	...	...	...	...	...	...
Oak, black.	Quercus velutina.	Tennessee.	Mixed hardwood, seedling.	4.8	35	7.8	52	12.0	61	14.2	68	18.3	78	22.0	85
		Kentucky.	Mixed hardwood, sprout.	6.9	53	11.3	75	13.8	80	14.4	81	...	...	...	...
Oak, chestnut.	Quercus prinus.	Tennessee.	Mixed hardwood, seedling.	3.3	20	5.6	35	8.8	45	11.0	53	16.0	66	21.0	76
Oak, red.	Quercus rubra.	Kentucky.	Mixed hardwood, sprout.	7.6	56	12.4	76	16.4	86	...	...	...	...	...	...
Oak, white.	Quercus alba.	Tennessee.	Mixed hardwood, seedling.	3.5	27	5.8	34	9.0	56	11.2	65	17.0	82	22.2	92
		Kentucky.	Mixed hardwood, sprout.	5.6	40	8.7	57	11.2	69	12.2	72	14.4	77	...	...
Walnut, black.*	Juglans nigra.	Illinois.	Farm plantations (moist soil).	8.0	40	12.0	60	...	...	...	...	...	...	...	...
Poplar, yellow.	Liriodendron tulipifera.	Tennessee.	Cove — virgin forest — Second growth in open.	5.0	39	8.7	60	13.7	81	17.0	90	24.0	100	28.2	105
		Virginia.	Quality I — seedling.	9.7	64	15.2	83	...	...	...	...	...	...	...	...

* These species were grown in plantations on farm land, and have a correspondingly faster growth than the other forest-grown species.

Note. — The diameter given is 4.5 feet from the ground.

From Report of the National Conservation Commission Rate of Forest Growth, by E. A. Zeigler, page 207.

TABLE IX. — APPROXIMATE AVERAGE RATE OF GROWTH FOR SOUTHERN FORESTS.

Species.		Locality.	Forest type.	Age.											
				30 years.		50 years.		80 years.		100 years.		150 years.		200 years.	
Common name.	Botanical name.			Diameter.	Height.	Diameter.	Height.	Diameter.	Height.	Diameter.	Height.	Diameter.	Height.	Diameter.	Height.
				In.	Ft.	In.	Ft.	In.	Ft.	In.	Ft.	In.	Ft.	In.	Ft.
Ash, white.	Fraxinus americana.	Arkansas.	Bottom land.	7.8	67	12.1	90	17.5	103	20.3	106	....	...	....	...
Cedar, eastern red.	Juniperus virginiana.	Alabama.	" Forest grown."	5.0	35	9.0	50	14.0	59	....	...	....	...	....	...
Cottonwood.	Populus deltoides.	Mississippi.	Bottom land (sprout?).	21.7	126	30.4	143	38.4	152	....	...	....	...	....	...
Cypress, bald.	Taxodium distichum.	Maryland.	Swamp.	3.9	31	7.8	53	15.2	82	20.5	95	35.0	107	....	...
Gum, black.	Nyssa sylvatica.	No data.	....	....	...	....	...	....	...	....	...	....	...	....	...
Gum, red.	Liquidambar styraciflua.	South Carolina.	Bottom land.	11.2	87	16.8	106	23.4	116	26.9	120	34.2	125	....	...
Pine, Cuban.	Pinus heterophylla.	South Carolina.	Lowland pine type.	6.2	44	10.4	67	15.0	86	17.3	93	21.7	102	25.0	105
Pine, loblolly.	Pinus taeda.	South Carolina.	Pine type.	10.0	59	15.9	86	21.5	104	24.5	111	30.7	122	....	...
Pine, longleaf.	Pinus palustris.	South Carolina.	Pine type.	3.0	29	6.7	52	12.4	78	15.5	86	20.0	95	23.6	100
Pine, scrub.	Pinus virginiana.	Maryland.	Pure stand — Dom.	6.5	46	8.9	63	....	...	....	...	....	...	....	...
Pine, shortleaf.	Pinus echinata.	Arkansas.	Pine type.	4.2	35	7.5	53	11.9	72	14.6	81	19.8	93	23.4	100

Note. — The diameter given is 4.5 feet from the ground.

From Report of the National Conservation Commission Rate of Forest Growth, by E. A. Zeigler, page 209.

TABLE X. — APPROXIMATE AVERAGE RATE OF GROWTH FOR ROCKY MOUNTAIN FORESTS.

Species.		Locality.	Forest type.	Age.											
				30 years.		50 years.		80 years.		100 years.		150 years.		200 years.	
Common name.	Botanical name.			Diameter.	Height.	Diameter.	Height.	Diameter.	Height.	Diameter.	Height.	Diameter.	Height.	Diameter.	Height.
				In.	Ft.	In.	Ft.	In.	Ft.	In.	Ft.	In.	Ft.	In.	Ft.
Fir, alpine.	Abies lasiocarpa.	No data.	....	....	...	....	...	....	...	....	...	....	...	....	...
Fir, Douglas	Pseudotsuga taxifolia.	Idaho.	Fir slope — Quality I.	1.6	8	5.9	35	12.3	63	15.6	72	21.1	86	24.9	93
Fir, white.	Abies concolor.	No data.	....	....	...	....	...	....	...	....	...	....	...	....	...
Larch, western	Larix occidentalis.	No data.	....	....	...	....	...	....	...	....	...	....	...	....	...
Pine, lodgepole.	Pinus contorta.	Montana.	Lodgepole slope.	4.0	40	6.6	60	9.6	70	11.2	73	14.0	76	....	...
Pine, silver.	Pinus monticola.	No data.	....	....	...	....	...	....	...	....	...	....	...	....	...
Pine, western yellow.	Pinus ponderosa.	Arizona.	Pure stand.	4.3	18	8.1	30	12.0	43	14.0	48	18.7	61	21.6	68
Spruce, Englemann.	Picea engelmanni.	Colorado.	Dry slope.	....	...	1.6	...	4.3	...	6.2	...	10.1	60	13.7	75

Note. — The diameter given is 4.5 feet from the ground.

From Report of the National Conservation Commission Rate of Forest Growth, by E. A. Zeigler, page 211.

TABLE XI. — APPROXIMATE AVERAGE RATE OF GROWTH FOR PACIFIC COAST FORESTS.

Species. Common name.	Species. Botanical name.	Locality.	Forest type.	Age 30 years. Diameter. In.	30 years. Height. Ft.	50 years. Diameter. In.	50 years. Height. Ft.	80 years. Diameter. In.	80 years. Height. Ft.	100 years. Diameter. In.	100 years. Height. Ft.	150 years. Diameter. In.	150 years. Height. Ft.	200 years. Diameter. In.	200 years. Height. Ft.
Cedar, western red.	Thuya plicata.	No data.	....	....	...	....	...	....	...	....	...	....	...	....	...
Fir, Douglas.*	Pseudotsuga taxifolia.	Washington.	Fir-hemlock (virgin).	7.4	48	13.7	80	20.5	118	24.3	138	31.3	180	37.4	208
			Fir-hemlock (second growth)	10.5	68	14.6	114	....	...	....	...	....	...	....	...
Fir, white.	Abies concolor.	California.	Plateau, slope and bottom	0.7	8	3.9	24	9.0	57	12.2	76	19.5	111	26.0	136
Gum, blue.†	Eucalyptus globulus.	Southern California.	Plantations — Quality I seedlings	28.0	140	....	...	....	...	....	...	....	...	....	...
Hemlock, western.	Tsuga heterophylla.	Washington.	Pure hemlock (virgin).	3.8	35	7.6	55	12.4	93	15.2	110	21.6	142	27.6	165
			Pure hemlock (second growth)	8.2	62	12.6	96	15.7	122	17.1	132	....	...	....	...
Pine, sugar.	Pinus lambertiana.	California.	Mixed slope (Sierras)	41.	22	7.8	39	14.4	70	18.7	92	28.2	134	32.4	144
Pine, Western yellow.	Pinus ponderosa.	California.	Mixed sugar pine	17.1	35	12.2	61	17.8	87	20.9	101	26.5	121	29.7	130
Redwood.	Sequoia sempervirens.	California.	Moist flat-sprout	10.7	71	14.5	95	....	...	....	...	....	...	....	...

* The diameter given is 4.5 feet from the ground, except for Douglas fir, which is for a stump height a little higher than 4.5 feet.

† An exotic introduced from Australia and widely planted for windbreak and fuel purposes in Southern California.

From Report of the National Conservation Commission Rate of Forest Growth, by E. A. Zeigler, page 213.

GLOSSARY OF TERMS IN FORESTRY AND LOGGING.

Largely taken from U. S. Forest Service Bulletin No. 61.

Accretion. Increase in diameter or height; distinguished from increment, increase in volume.

Accretion borer. An instrument for determining the growth in diameter of standing trees. It consists of a hollow auger, which, when bored into a tree, extracts a section showing the annual rings.

Accretion thinning. A thinning made specifically to increase the rate of growth in diameter of the trees which are left standing.

Advance growth. Young trees which have sprung up in accidental openings in the forest or under the forest cover before reproduction cuttings are begun.

After growth. Young trees which have sprung up as the result of reproduction cuttings.

Angle mirror. An instrument for turning of angles in subdividing land.

Annual ring. The layer of wood produced by the diameter growth of a tree in one year, as seen on a cross section.

Arboriculture. The growing of trees singly or in groups for any purpose.

Aspect. The direction toward which a slope faces. The eight main points of the compass, N., N.E., E., S.E., S., S.W., W., N.W., are distinguished in forest description.

Assimilation. In plants, the production of organic matter from inorganic matter.

Back fire. A fire started purposely some distance ahead of a fire which is to be fought. The back fire is intended to burn only against the wind, so that when the two fires meet, both must go out for lack of fuel.

Ball planting. A method of transplanting young trees with balls or lumps of earth around the roots.

Barn boss. One who has charge of the stables in a logging camp.

Basal area. The area of a cross section of a tree, or the sum of such areas.

Bast. The woody fibrous tissue of the inner bark.

Bed a tree, to. To level up the path in which a tree is to fall, so that it may not be shattered.

Binding chain. A chain used to bind together a load of logs.

Birl, v. To cause a floating log to rotate rapidly by treading upon it.

Blank. An opening in the forest where, from any cause, very few or no trees are growing.

Blaze, v. To mark, by cutting into trees, the course of a boundary, road, trail or the like.

Board foot. The contents of a board one foot square and one inch thick. The common unit of measure for logs and lumber in the United States.

Board measure. The standard of lumber measurement, the unit of which is the board foot.

Board rule. A graduated stick for determining the contents of boards. The number of board feet in boards of given widths and lengths is shown upon the stick.

Body wood. Cord wood cut from those portions of the stems of trees which are clear of branches.

Boom, n. Logs or timbers fastened together end to end and used to hold floating logs. The term sometimes includes the logs inclosed, as a boom of logs.

Boom company. A corporation engaged in handling floating logs, and owning booms and booming privileges.

Brake sled. A logging sled so constructed that, when the pole team holds back, a heavy iron on the side of each runner of the forward sled is forced into the roadbed.

Breasthigh. At or having a height of 4½ feet above the ground.

Broad-leaved trees. Applies to trees whose leaves have a broad flat surface, unlike the needle- or awl-shaped leaves of the conifers.

Buck, v. To saw felled trees into logs.

Bull chain. 1. A very heavy chain, to which a number of short chains, with hooks on one end and dogs on the other, are attached. It is used to draw logs from the mill pond up the gangway.

2. *See* Jack chain.

Bull donkey. A large donkey engine which, by drum and cable, drags logs from the place where they are yarded to a landing.

Bummer, n. A small truck with two low wheels and a long pole, used in skidding logs.

Bunch logs, to. To collect logs in one place for loading.

Bunk, n. 1. The heavy timber upon which the logs rest on a logging sled.

2. The cross beam on a log car or truck, on which the logs rest.

3. A log car or truck.

4. A bed in a logging camp.

Burn. An area over which fire has run to the noticeable injury of the forest.

Butt off, to. 1. To cut a piece from the end of a log on account of a defect.

2. To square the end of a log.

Caliper. An instrument for measuring the diameter of trees or logs, usually consisting of a graduated beam to which is attached one fixed and one sliding arm.

Cambium. In trees and shrubs, the layer of new growing tissue between the bark and wood.

Cant hook. A tool like a peavey, but having a toe ring and lip at the end instead of a pike.

Catamaran, n. A small raft carrying a windlass and grapple, used to recover sunken logs.

Catface, n. A partly healed over fire scar on the stem of a tree.

Clean cutting. 1. The cutting of the entire stand.

2. An area upon which the entire stand has been cut.

Clean cutting method. A method of conservative lumbering in which the entire stand is cut at one time and reproduction is secured by sowing or planting.

Cleaning. A thinning made in a stand which has not reached the small-pole stage. Its main object is to remove trees of undesirable form and species.

Clear length. In silvics, that portion of the stem of a tree free from branches. In forest measurements the meaning of the term varies with the species measured and the purpose of the measurements. For example, clear length is in some cases used to designate that portion of the stem entirely free from branches, in others that portion free from dead branches, or from growing branches of a given size.

Coal off, to. To cut a forest clean for charcoal wood.

Compartment. The unit of area treated in the working plan. The size and the shape of compartments are determined mainly by topographic features.

If a compartment contains a stand varying greatly in composition, age, or needs, it may be divided into two or more subcompartments, which may be either temporary or permanent.

Compartment line. The boundary of a compartment. It may be marked by a road, a ride, or a natural feature, such as a stream or the crest of a ridge. In Europe, when other demarcation is wanting, clean cuttings upon narrow strips are made to mark the boundaries of a compartment. These are known as rides. A ride which separates two cutting series, and thus runs parallel to the prevailing wind direction, is called a major ride, while one which completes the demarcation of a compartment is known as a minor ride.

Composite forest. A forest in which both seedlings and sprouts occur in considerable number. It may be either pure or mixed.

Conifer. A member of the Pine Family or Coniferæ.

Coniferous. Cone-bearing.

Conk, n. 1. The decay in the wood of trees caused by a fungus.

2. The visible fruiting organ of a tree fungus.

Conservative lumbering. Practical forestry; any method of lumbering which perpetuates the forest by use.

Conversion. A change from one system or method of forest management to another, as from the sprout system to the seed system.

Conversion period. The period during which the change from one system or method of forest management to another is effected.

Cookee, n. Assistant cook and dishwasher in a logging camp.

Coppice. A forest grown from sprouts.

Crown. In silvics, the upper part of a tree, including the living branches with their foliage. In forest measurements the use of the term varies with the kind of tree and the purpose of the measurements. For example, crown may be used to designate the whole leaf and branch system, or that portion of it above a dead or a growing branch of a given size. In tree description the crown is described as long or short, broad or narrow, compact or ragged, conical or flat.

Crown density. The density of the crowns of the trees in a forest; it is usually measured by the extent to which the ground is shaded. The degrees of crown density in a forest and expressed by the following terms: closed, — when the crowns form an uninterrupted cover and permit little or no sunlight to reach the ground; dense, — when three-fourths or more of the ground is shaded by the crowns; thin, — when three-fourths to one-half of the ground is shaded by the crowns; open, — when less than one-half the ground is shaded by the crowns. Park forest is forest in which shade occurs only in isolated patches, under single trees or small groups.

Cruise, v. To estimate the amount and value of standing timber.

Cruiser, n. One who cruises.

Cull. 1. *v.* To take out of a forest by selection a portion of the trees.
2. *n.* A low-grade log or piece of lumber.

Culled forest. Forest from which cuttings by selection have removed a portion of the trees.

Current annual increment. The volume of wood produced in a given year by the growth of a tree or stand.

Cut, n. A season's output of logs.

Cut over, to. To cut most or all of the merchantable timber in a forest.

Cut-over forest. Forest in which most or all of the merchantable timber has been cut.

Cutting. A piece of a leaf, branch, stem, or root which when inserted in moist material is capable of sending out roots and forming a new plant; a slip.

Cutting area. The area over which cuttings are to be or have been made.

Cutting height. The height above the ground at which a tree is to be cut.

Cutting series. A block or a part of a block containing even-aged stands whose ages differ uniformly within given limits and which are to be cut in turn, the cuttings usually following a given direction. A perfect cutting series seldom exists, except under the clean-cutting method followed by artificial reproduction, or under the sprout method.

Deaden. To kill a standing tree by girdling it.

Deadening. An area upon which the trees have been deadened.

Deadhead, n. A sunken or partly sunken log.

Delinquent tax lands. Lands on which taxes have not been paid. They are offered for sale at stated times after public notice, and tracts which find no buyers revert to the state.

Diameter, breasthigh. The diameter of a tree at $4\frac{1}{2}$ feet above the ground.

Diameter class. All trees in a stand whose diameters are within prescribed limits.

Diameter growth. The increase in diameter of a tree.

Diameter limit. The diameter, usually breasthigh, which defines the size to which trees are to be measured or used for any given purpose.

Diameter tape. A tape for ascertaining the diameter of trees, so graduated that the diameter corresponding to the girth of a tree is read directly from the tape.

Dibble. A tool for making holes for planting seeds or young trees.

Dibble in, to. To plant seeds or young trees in holes made with a dibble.

Diœcious. Staminate and pistillate flowers borne on different plants.

Dominant. Having the crown free to light on all sides because of greater height.

Doty, a. Decayed.

Dray, n. A single sled used in dragging logs. One end of the log rests upon the sled.

Drive, n. 1. A body of logs or timber in process of being floated from the forest to the mill or shipping point.

2. That part of logging which consists in floating logs or timbers.

Dry roll, to. In sacking the rear, to roll stranded logs into the bed of the stream from which the water has been cut off preparatory to flooding.

Dry topped. Having a dead or a partially defoliated crown, or discolored foliage, as the result of injury or disease.

Duffle, n. The personal belongings of a woodsman or lumberjack which he takes into the woods.

Experiment area. A forest area of known size upon which successive measurements or other detailed studies are made for the determination of the growth and behavior of the stand, or upon which experiments are conducted to ascertain the effect of methods of treatment upon the forest.

Fail spot. A place where natural or artificial reproduction has failed.

Faller, n. One who fells trees.

Falling ax. An ax with a long helve and a long, narrow bit, designed especially for felling trees.

False ring. The layer of wood, less than a full season's growth, and seldom extending around the stem, which is formed whenever the diameter growth of a tree is interrupted and begins again during the same growing season.

Filer, n. One who files the crosscut saws in the woods.

Final yield. All material derived from reproduction cuttings or clean cuttings. It is usually the chief crop, and marks the end of the rotation.

Firebreak. An opening, ploughed strip of land or anything which prevents the spread of fires in the forests or elsewhere.

Fire line. A strip kept clear of inflammable material as a protection against the spread of forest fire.

First growth. 1. Natural forest in which no cuttings have been made.

2. Trees grown before lumbering or severe fire entered the forest; belonging to the original stand.

Fitter, n. One who notches the tree for felling and after it is felled marks the log lengths into which it is to be cut.

Flume, n. An inclined trough in which water runs, used in transporting logs or timbers.

Forest, n. An area whose principal crop is trees. A forest includes both the forest cover and the soil beneath it. A forest judged by the character of the stand may be timberland or woodland. These constitute the two great classes of forest, between which it is possible to draw a practical but not an absolute distinction. Timberland may be broadly defined as that class of forest which contains in commercial quantities trees of sufficient size and of the required kind to furnish saw logs, pulp wood, ties, poles or wood for similar uses.

Woodland may be broadly defined as forest which contains trees fit for firewood or fencing but none or very few trees which are suitable for the uses enumerated above.

A timber tract is a body of timberland, usually of large area.

A woodlot is a forest of small area in which the wood is used mainly for fuel, fencing, and other farm purposes.

Forest capital. The capital which a forest represents. It consists of the forest land, or fixed capital, and the stand.

Forest cover. All trees and other plants in a forest.

Forester. One who practices forestry as a profession.

Forest expectation value. The present net value of all future returns expected from the forest capital. It is determined by discounting to the present time, at compound interest, all returns and expenses anticipated.

Forest extension. The establishment of forest upon areas where it is at present absent or insufficient.

Forest fire. A fire in timberland or woodland. A forest fire may be a ground fire, a surface fire, a stand fire or a crown fire. A ground fire is one which burns in the forest floor and does not appear above the ground. When a fire runs over the surface or burns the undergrowth, it is a surface fire. When a surface fire spreads from the undergrowth to the stand, igniting the trees, it becomes a stand fire. Under certain conditions the crowns of the trees may be ignited, causing a crown fire.

Forest floor. The deposit of vegetable matter on the ground in a forest. Litter includes the upper, but slightly decomposed portion of the forest floor; humus, the portion in which decomposition is well advanced.

Forest grown. Grown in the forest from self-sown seed.

Forest influences. All effects resulting from the presence of the forest, upon health, climate (including wind, rainfall, temperature, etc.), stream flow, and economic conditions.

Forest management. The practical application of the principles of forestry to a forest area. Forest management includes: forest mensuration, or the determination of the present and future product of the forest; forest organization, or the preparation of working plans and planting plans, detailed and comprehensive schemes for the establishment and best use of the forest; and forest finance, or the determination of the money returns from forestry.

Three great systems of forest management are distinguished: the seed system, the sprout system, and the composite system. The seed system includes the stand method, group method, strip method, patch method, strip stand method, group seed method, scattered seed method, single tree method, reserve seed method, clean cutting method. The sprout system includes the sprout method. The composite system includes the reserve sprout method.

Forest nursery. An area upon which young trees are grown for forest planting.

Forest plantation. Forest growth, established by setting out young trees or by sowing seed, which has not reached the small-pole stage. A forest plantation, made by setting out young trees, which has passed the small-pole stage, is called a planted forest. A sown forest plantation which has passed the small-pole stage is called a sown forest.

Forest policy. The principles which govern the administration of the forest for its best permanent use.

Forest products. All usable material yielded by the forest. The following classes are distinguished: major products include all wood harvested for any purpose; minor products include all forest products except wood.

Forest protection. The safe guarding of the forest against any damage not caused by its own growth.

Forestral. Pertaining to forestry.

Forest replacement. The restoration of forest growth on denuded areas.

Forestry. The science and art of making the best permanent use of the forest. The main branches of forestry are forest policy, silviculture, forest management, forest protection and forest utilization.

Forest type. A forest or a part of a forest possessing distinctive characteristics of composition or habit of growth.

Forest utilization. The most profitable use of forest products, including lumbering, the various wood-using industries such as the wood pulp, wood tannin, cooperage, veneer, excelsior, and similar industries and the uses to which our woods are put.

Form class. All trees in a stand so similar in form that the same form factor is applicable in determining their actual volume.

Form factor. The ratio, expressed decimally, between the volume of a tree, or portion of a tree, and of a cylinder of the same height and diameter. The volume of this cylinder multiplied by the form factor gives the actual volume of the tree or portion of the tree. The three kinds of form factors are distinguished, according to the portion of the tree to which they refer. A tree form factor is used for determining the actual volume of the whole tree; a stem form factor for determining the volume of the stem; and a timber form factor for determining the merchantable contents of stem, crown, or both.

A form factor is called absolute when the diameter of the tree is measured at any convenient height, the form factor referring only to that portion of the tree above the point at which the diameter is measured; normal, when the diameter is measured at a height in constant ratio to the total height of the tree; and artificial, when the breasthigh diameter is measured.

Full scale. Measurement of logs, in which no reduction is made for defects.

Future yield. The amount of wood which given trees upon a given area will contain after a given period.

Future yield table. A tabular statement of future yield.

Germination. The process by which a seed or spore gives rise to a new and independent plant.

Girdling. The act of cutting through the inner bark and sapwood to cut off the circulation of the sap. Practised in the Southern Appalachians as a means of quickly clearing agricultural land.

Ground cover. All small plants growing in a forest, except young trees; such as ferns, mosses, grasses and weeds.

Group method. A method of conservative lumbering in which groups of young trees which have sprung up in openings caused by logging, insect damage, windfall, snowbreak or other agency, are taken as starting points for the future forest; or if these are insufficient, small openings are purposely made. Reproduction by self-sown seed from the mature stand at the edges of these groups is secured by careful cuttings, which extend the groups until they join.

Group mixture. A mixed forest in which trees of the same species occur in groups not large enough to be considered pure stands.

Group seed method. A method of conservative lumbering in which the forest is reproduced after a single cutting, by leaving in groups seed trees of the kind desired.

Harden off, to. To prepare seedlings in the seedbed for transplanting by gradually exposing them to wind and sunlight.

Hardwood, n. A broadleafed, or dicotyledonous, tree.

Haul, n. In logging, the distance and route over which teams must go between two given points, as between the yard or skidway and the landing.

Head driver. An expert river driver who during the drive is stationed at a point where a jam is feared. Head drivers usually work in pairs.

Head faller. The chief of a crew of fallers.

Heel in, to. To store young trees for planting by laying them against the side of a trench and covering the roots with earth.

Height class. All trees in a stand whose heights are within prescribed limits.

Height growth. The increase in height of a tree.

Height measure. An instrument for measuring the height of a tree.

Humus. Decomposed organic matter in and on the surface of the soil.

Hypsometer. An instrument for taking heights of trees.

Ice a road, to. To sprinkle water on a logging road so that a coating of ice may form, thus facilitating the hauling of logs.

Improvement thinning. Usually the first thinning made when a forest is put under management, to prepare it for the application of a regular system.

Increment. The volume or value of wood produced during a given period by the growth of a tree or of a stand. Three kinds of increment are distinguished: volume increment is the increase in volume of a tree or stand; quality increment is the increase in value per unit of volume; price increment is the increase resulting from an increase in the price of forest products independent of quality increment.

Index. The highest average actually found upon a given locality. The term index applies to stand, diameter growth, height growth, increment, and present and future yield is the equivalent of normal, when normal is used to describe the assumed standard based upon actual measurement.

Index forest. That forest which in density, volume and increment reaches the highest average which has been found upon a given locality. Measurements of such a forest provide a standard for comparison with other forests of the same age and composition, grown under similar conditions.

Intermediate. Having the crown shaped on the sides, but free to light at the top.

Intermediate yield. All material from thinnings or from any cutting not intended to invite or assist reproduction.

Intolerant. Incapable of enduring heavy shade.

Irregular forest. Forest in which the trees differ considerably in age.

Jack chain. An endless spiked chain, which moves logs from one point to another, usually from the mill pond into the sawmill.

Jam, n. A stoppage or congestion of logs in a stream, due to an obstruction or to low water.

Jam, to break a. To start in motion logs which have jammed.

Key log. In river driving, a log which is so caught or wedged that a jam is formed and held.

Landing, n. 1. A place to which logs are hauled or skidded preparatory to transportation by water or rail. A rough and tumble landing is one in which no attempt is made to pile logs regularly.

2. A platform, usually at the foot of a skid road, where logs are collected and loaded on cars. A lightning landing is one having such an incline that the logs may roll upon the car without assistance.

Large-pole forest. A forest of large poles.

Large-sapling forest. A forest of large saplings.

Layer. A shoot which, while attached to the plant, takes root at one or more places and forms a new plant.

Lift. To pry up seedlings in the seedbed, so that they may be pulled up by hand for transplanting.

Line out, to. To transplant seedlings from the seedbed to rows in the forest nursery.

Litter. That portion of the forest floor which is not in an advanced state of decomposition.

Loam. Friable, mellow, rich soil containing much humus.

Locality. An area, considered with reference to forest-producing power; the factors of the locality are the altitude, soil, slope, aspect and other local conditions influencing forest growth.

Locality class. All localities with similar forest-producing power.

Log, to. To cut logs and deliver them at a place from which they can be transported by water or rail, or, less frequently, at the mill.

Log rule. 1. A tabular statement of the amount of lumber which can be sawed from logs of given lengths and diameters.

2. A graduated stick for measuring the diameter of logs. The number of board feet in logs of given diameters and lengths is shown upon the stick.

Logging sled. The heavy double sled used to haul logs from the skidway or yard to the landing.

Lumber, v. To log, or to manufacture logs into lumber, or both.

Lumberjack, n. One who works in a logging camp.

Many-aged forest. A forest through all parts of which many different age classes of trees tend to distribute themselves. When all age classes are thus distributed, the forest is all aged. These two terms replace selection forest, many-aged being substituted for imperfect selection, and all aged for perfect or ideal selection.

Market, n. A log 19 inches in diameter at the small end and 13 feet long.

Marking hatchet. A hatchet for marking trees. A raised die is cut on the head for stamping the face of the blaze.

Mature forest. Forest so old that growth in height is practically at an end, and diameter growth is decreasing.

Mean annual increment. The total increment of a tree or stand divided by its age in years.

Merchantable length. The total length of that portion of the stem which can be used under given conditions.

Merchantable volume. The total volume of that portion of the tree which can be used under given conditions.

Mild humus. Humus in a condition favorable to forest growth.

Mixed forest. Forest composed of two or more species.

Monœcious. Both staminate and pistillate flowers borne on the same plant (e.g., black walnut).

Mound planting. A method of planting on wet ground, in which the seeds or young trees are planted on mounds, ridges or hills.

Mulch. Any loose material that protects the soil from frost or evaporation.

Muskeg. A term commonly applied to sphagnum swamps by the Indians and woodsmen of the Northern States.

National forest. A forest which is the property of the United States.

National forest reserve. A tract of land set apart from the public domain by proclamation of the President under section 24 of the act of March 3, 1891, or created by special act of Congress, and administered under laws of the United States passed for that purpose, in order "to improve and protect the forest within the reservation, or for the purpose of securing favorable conditions of waterflows and to furnish a continuous supply of timber for the use and necessities of citizens of the United States."

National park. A tract of Government land withdrawn by special act of Congress from settlement, occupancy or sale, under the laws of the United States, for the benefit and enjoyment of the people.

Nurse. A tree which fosters the growth of another in youth.

Nursery. An establishment for the raising of plants.

Nursery grown. Grown in a forest nursery.

Open grown. Said of trees when not grown sufficiently close to have their crowns meet.

Osiers. A class of willows used for baskets.

Overmature forest. Forest in which, as the result of age, growth has almost entirely ceased, and decay and deterioration have begun.

Overtopped. Having the crown shaded from above, although a side or sides may be free to light.

Parasite. A plant or animal that lives upon and obtains its food from other living plants or animals.

Patch method. The clean cutting of small patches to invite reproduction by self-sown seed from the surrounding forest.

Patch sowing. Sowing forest seed in spots.

Peavey, n. A stout lever 5 to 7 feet long, fitted at the larger end with a metal socket and pike, and a curved steel hook which works on a bolt; used in handling logs, especially in driving. A peavey differs from a cant hook in having a pike instead of a toe ring and lip at the end.

Pecky, a. A term applied to an unsoundness most common in bald cypress.

Periodic annual increment. The total increment for the period, divided by the number of years in the period.

Periodic increment. The volume of wood produced by the growth of a tree or stand in a specified number of years.

Pike pole. A piked pole, 12 to 20 feet long, used in river driving.

Pitch pocket. A cavity in wood filled with resin.

Pitch streak. A seam or shake filled with resin.

Planting plan. A detailed scheme for forest planting on a given area.

Planting site. An area which is to be artificially stocked with forest growth.

Pole. A tree from 4 to 12 inches in diameter, breasthigh.

A small pole is a tree from 4 to 8 inches in diameter, breasthigh.

A large pole is a tree from 8 to 12 inches in diameter, breasthigh.

Pollard, v. To invite the production of shoots at the top of a tree by cutting back the crown.

Preliminary examination. A reconnoissance of a forest to determine whether the preparation of a working plan for its management is advisable, or a reconnoissance to determine the advisability of forest planting.

Present yield. The amount of wood at present contained in given trees upon a given area.

Present yield table. A tabular statement of the amount of wood at present contained in given trees upon a given area.

Private forest. A forest which is the property of an individual, corporation, company or private institution.

Protection forest. A forest whose chief value is to regulate stream flow, prevent erosion, hold shifting sand or exert any other indirect beneficial effect.

Pruning. The removal of branches from standing trees by natural or artificial means. The clearing of the stem through the death and fall of branches for want of light is known as natural pruning. When living branches are removed by cutting them close to the stem the operation is known as green pruning; when it is confined to dead branches, as dry pruning.

Puddle, v. To dip the roots of young trees in thin mud.

Puddle, n. A mixture of soil or mold and water, forming thin mud, in which the roots of young trees are dipped to retard drying out during transplanting.

Pure forest. Forest composed of trees of one species. In practice, a forest in which 80 per cent of the trees are of one species.

Quincunx planting. A method of planting in which young trees are set in the center, and at each corner of successive squares.

Regular forest. Forest in which the trees are approximately of the same age.

Reproduction. 1. The process by which a forest is renewed. Natural reproduction is the renewal of a forest by self-sown seeds, or by sprouts. Artificial reproduction is the renewal of a forest by sowing or planting.

2. Seedlings or saplings from sprouts or from self-sown seed.

Reproduction cutting. Any cutting intended to invite or assist reproduction.

Reproduction period. The space of time required for the renewal of a stand.

Reserve seed method. That method of conservative lumbering in which, in a stand which is being reproduced by self-sown seed, a number of trees are left uncut for a period, usually a second rotation, after the stand itself is reproduced.

Reserve sprout forest. Two-storied forest, in which sprouts form the lower, and seedlings, or selected, healthy sprouts, the upper story.

Reserve sprout method. That method of conservative lumbering in which an overwood composed of seedling trees, or selected sprouts, is maintained above a stand of sprouts.

Restock. To renew a forest, either by natural or artificial means.

Rise, n. The difference in diameter, or taper, between two points in a log.

River boss. The foreman in charge of a log drive.

Rock. In forest description rock refers to those characteristics of the underlying formation which affect the forest; as, for example, its outcrop, composition, and the rapidity of its disintegration.

Rock in, to. To plant young trees in openings in the ground made by prying or rocking a spade back and forth.

Root. A part of the plant which absorbs nourishment for the plant, or serves as a support.

Root collar. That place at the base of a tree where the swelling which is the direct result of the ramifications of the roots begins.

Rotation. The period represented by the age of a forest, or a part of a forest, at the time when it is cut, or intended to be cut. The following classes of rotation are distinguished: financial rotation, under which a forest yields the highest net interest on its capital value, calculating at compound interest; income rotation, under which a forest yields the highest net return calculating without interest; silvical rotation, the rotation most favorable to the natural reproduction of the forest under a given method; technical rotation, under which a forest yields the material most suitable for a certain purpose; volume rotation, under which a forest yields the greatest quantity of material.

Row planting. A method of planting in which the young trees are placed in rows, the distance between the rows being greater than the distance between the young trees in the rows. In planting seeds or seedlings in the forest nursery this method is known as drill planting.

Sample tree. A tree which in diameter, height and volume is representative of a tree class. A class sample tree is a tree which in diameter, height and volume represents the average of several tree classes.

Sapling. A tree 3 feet or over in height, and less than 4 inches in diameter, breasthigh. A small sapling is a sapling from 3 to 10 feet in height. A large sapling is a sapling 10 feet or over in height.

Scattered seed method. That method of conservative lumbering in which reproduction is provided for by leaving, after a single cutting, scattered seed trees of the kind desired.

Scratcher. An instrument used for marking trees. It usually consists of a hook-like gouge fastened to a flat, elliptical iron hoop, with wooden handle plates on the opposite side from the gouge.

Scaler, n. One who determines the volume in logs.

Second growth. Forest growth which comes up naturally after cutting, fire or other disturbing cause.

Seed. The ripened ovule.

Seedbed. A specially prepared area, usually in the forest nursery, for the raising of seedlings.

Seed forest. A forest composed wholly or mainly of trees from seed.

Seedling. 1. A tree grown from seed.
2. A tree grown from seed which has not reached a height of 3 feet.

Seed spot. A small area, usually in a burn or in an opening in the forest, which is sown with tree seed.

Seed system. One of the three great systems of forest management. Under it, reproduction is obtained from seed.

Seed tree. Any tree which bears seed; specifically, a tree which provides the seed for natural reproduction.

Seed year. A year in which a given species of tree bears seed; specifically, a year in which a given species bears seed abundantly.

Selection forest. See many-aged forest.

Self-sown seed. Strictly, disseminated without the intervention of human or animal agency; in common practice, seed sown by any agency other than man.

Semi-mature forest. Forest in which rapid growth in height has culminated, but diameter growth has not begun to fall off.

Setting, n. The temporary station of a portable sawmill, a yarding engine or other machine used in logging.

Severance cutting. The cutting of all trees upon a narrow strip before natural pruning has far advanced, in order that the trees bordering this strip may, as the result of partial exposure, become wind-firm through the development of strong roots. Thus severance cuttings are made to strengthen the trees on the edge of a stand which will later be entirely exposed through the removal of the stand which now protects it.

Shade frame. A frame for the partial shading of a seedbed. It consists of a cover of laths, brush or cloth, supported on posts, and arranged so that light can be admitted as desired.

Shake, n. A crack in timber, due to frost or wind.

Shelterbelt. Natural or artificial forest maintained as a protection from wind or snow. A narrow shelterbelt in which true forest conditions do not exist is a windbreak when maintained as a protection against wind, and a snowbreak when maintained as a protection against snow.

Shrub. A woody plant with no main stem or trunk.

Silvical. Pertaining to silvics.

Silvics. 1. The science which treats of the life of trees in the forest.

2. The habit of behavior of a tree in the forest.

Silviculture. The art of producing and tending a forest; the application of the knowledge of silvics in the treatment of a forest.

Single-tree method. That method of conservative lumbering in which reproduction from self-sown seed under the shelter of the old stand is invited by the cutting of single trees. This cutting may be made throughout the forest, as in some woodlots, or in definite portions of the forest in turn.

Single-tree mixture. A mixture in which trees of different species occur singly.

Skid, n. A log or pole, commonly used in pairs, upon which logs are handled or piled.

Skid v. 1. To draw logs from the stump to the skidway, landing, or mill.

2. As applied to a road, to reënforce by placing logs or poles across it.

Skidder, n. 1. One who skids logs.

2. A steam engine, usually operating from a railroad track, which skids logs by means of a cable.

3. The foreman of a crew which constructs skid roads.

4. *See* Bummer.

Skid road. 1. A road or trail leading from the stump to the skidway or landing.

2. A road over which logs are dragged, having heavy transverse skids partially sunk in the ground, usually at intervals of about 5 feet.

Skidway, n. Two skids laid parallel at right angles to a road, usually raised above the ground at the end nearest the road. Logs are usually piled upon a skidway as they are brought from the stump for loading upon sleds, wagons or cars.

Slash, n. 1. The debris left after logging, wind or fire.

2. Forest land which has been logged off and upon which the limbs and tops remain, or which is deep in debris as the result of fire or wind.

Slide, n. A trough built of logs or timber, used to transport logs down a slope.

Slope. The gradient of the land surface. In forest description the following terms are used to define the slope, each of which has its equivalent in percentages of the horizontal distance and in degrees:

	Per Cent	Degrees
Level	0– 5	0.0– 3.0
Gentle	5– 15	3.0– 8.5
Moderate	15– 30	8.5–16.5
Steep	30– 50	16.5–26.5
Very steep	50–100	26.5–45.0
Precipitous	over 100	over 45.0

Small-pole forest. A forest of small poles.

Small-sapling forest. A forest of small saplings.

Snowbreak. 1. The breaking of trees by snow.

2. An area on which trees have been broken by snow.

Snub, v. To check, usually by means of a snub line, the speed of logging sleds or logs on steep slopes, or of a log raft.

Soil. In forest description the origin, composition, depth and moisture of the forest soil are considered under soil. Its depth is defined by the following terms, each of which has its equivalent in inches.

Very shallow	less than 6 inches.
Shallow	6 to 12 inches.
Moderate	12 to 24 inches.
Deep	24 to 36 inches.
Very deep	over 36 inches.

The moisture of the soil is defined by the following terms: Wet — when water drips from a piece held in the hand without pressing; moist — when water drips from a piece pressed in the hand; fresh — when no water drips from a piece pressed in the hand, although it is unmistakably present; dry — when there is little or no trace of water; very dry — when the soil is parched. Such soils are usually caked and very hard, sand being an exception.

Sour humus. Humus harmful to forest growth owing to the presence of humic or similar acids produced by decomposition under excess of moisture and lack of air.

Species. A division of a genus, the plants of which seem to be derived from an immediate common ancestor.

Splash dam. A dam built to store a head of water for driving logs.

Spring board. A short board, shod at one end with an iron calk, which is inserted in a notch cut in a tree, on which the faller stands while felling the tree.

Sprinkler, n. A large wooden tank from which water is sprinkled over logging roads during freezing weather in order to ice the surface.

Sprout. A tree which has grown from a stump or root.

A shoot is a sprout which has not reached a height of 3 feet.

Sprout forest. A forest consisting wholly or mainly of sprouts.

Sprout method. That method of conservative lumbering in which reproduction is obtained by sprouts.

Sprout system. One of the three great systems of forest management, in which reproduction is secured by sprouts.

Spud, n. A tool for removing bark.

Square planting. A method of planting in which the distance between the rows is equal to the distance between the young trees in the rows.

Stand. All growing trees in a forest or in part of a forest.

Standard. A tree from 1 to 2 feet in diameter, breasthigh.

Standard forest. A forest of standards.

Stand class. All stands of similar density, height and volume for a given age or diameter and a given locality class. The index stand may constitute the first stand class.

Stand method. That method of conservative lumbering in which reproduction is secured from self-sown seed by means of successive cuttings made throughout the mature stand, thus leading to the production of a new stand approximately even aged. These successive cuttings encourage seed production, create conditions favorable to the growth of seedlings, and gradually remove the remaining trees of the mature stand as the young growth develops. The series of cuttings, which vary in number and duration according to the degree of difficulty with which reproduction is effected, is divided into the following four kinds: preparatory cuttings fit the stand for its reproduction by the removal of dead, dying or defective trees, and prepare the ground for the germination of seeds. A stand in which one or more preparatory cuttings have been made is in the preparatory stage. Seed Cuttings encourage seed production by the further opening of the stand, and admit light in quantity favorable for the development of young growth. A stand in which one or more seed cuttings have been made is in the seeding stage. Removal Cuttings gradually remove the mature stand which would otherwise retard the development of the young trees. A stand in which one or more removal cuttings have been made is in the removal stage. The final cutting is the last of the removal cuttings, in which all of the old stand still remaining is cut.

Stand table. A tabular statement of the number of trees of each species and diameter class upon a given area.

State forest. A forest which is the property of a state.

Stem. The trunk of a tree. The stem may extend to the top of the tree, as in some conifers, or it may be lost in the ramification of the crown, as in most broadleaf trees. In tree description the stem is described as long or short, straight or crooked, cylindrical or tapering, smooth or knotty.

Stem density. The extent to which the total number of trees in a given forest approaches the total number which the index forest of the same age and composition contains. It is ordinarily expressed as a decimal, 1 being taken as the numerical equivalent of the stem density of the index forest.

Storage boom. A strong boom used to hold logs in storage at a sawmill.

Stratification. A method of storing seeds with alternate layers of sand.

Stratify. To preserve tree seeds by spreading them in layers alternating with layers of earth or sand.

Strip method. That method of conservative lumbering in which reproduction is secured on clean-cut strips by self-sown seed from the adjoining forest.

Strip stand method. A modification of the stand method in which reproduction cuttings are not made simultaneously throughout the stand, but the stand is treated in narrow strips at such intervals that reproduction cuttings are generally going on in three strips at one time, one strip being in the removal stage, one in the seeding stage and one in the preparatory stage.

Stub. That portion of the stem left standing when a tree is accidentally broken off.

Stump. That portion of the tree below the cut made in felling a tree.

Stumpage, n. The value of timber as it stands uncut in the woods; or, in a general sense, the standing timber itself.

Stump age. The age of a tree as determined by the number of annual rings upon the face of the stump, without allowance for the period required for the growth of the tree to the height of the stump.

Stump height. The distance from the ground to the top of the stump, or from the root collar when the ground level has been disturbed. On a slope the average distance is taken as the stump height.

Sucker. A shoot from an underground root or stem.

Sun scald. An injury to the cambium caused by sudden exposure of a tree to strong sunlight.

Suppressed. Having growth more or less seriously retarded by shade.

Swamp, v. To clear the ground of underbrush, fallen trees and other obstructions preparatory to constructing a logging road or opening out a gutter road.

Swamper, n. One who swamps.

Swell butted. As applied to a tree, greatly enlarged at the base.

Tally board. A thin, smooth board used by a scaler to record the number or volume of logs.

Tally man. One who records or tallies the measurements of logs as they are called by the scaler.

Tangential sawing. The common way of cutting logs by which boards on each side of the center board are sawed by a cut that is tangent to the annual rings. This method serves to bring out the grain of wood most conspicuously.

Tap-root. A central root running deep into the soil.

Tensile strength. The force which resists breaking or drawing asunder.

Thinning. The removal of a portion of the trees with the object of improving the stand without inviting natural reproduction. The following kinds of thinnings are distinguished: cleaning, improvement thinning, accretion thinning.

Tolerance. The capacity of a tree to endure shade.

Tolerant. Capable of enduring more or less heavy shade.

Total increment. The total volume of wood produced by the growth of a tree or stand up to the time it is cut.

Tote, n. To haul supplies to a logging camp.

Town forest. A forest which is the property of a city, town or village.

Transpiration. The process by which water is taken up by the roots of plants and given off to the air through the leaves and branches.

Transplant. 1. *n.* A seedling which has been transplanted once or several times.

2. *v.* To take up a young tree and set it out again in another place.

Trap tree. A tree deadened or felled at a time when destructive bark beetles will be attracted to it and enter the bark. After they have entered, the bark is peeled and exposed to the sun, burned or buried, as the case may require, to destroy the insect.

Tree. A perennial woody plant with a single stem which from natural tendencies generally divides into two or more branches at some distance from the ground.

Tree analysis. A series of measurements and observations upon a felled tree to determine its growth and life history. Tree analyses vary with their purpose, and may include all or a part of the following, or may require additions to meet special needs. The usual measurements comprise length of each section, the diameter inside and outside the bark, the total age, the age and width of the sapwood, the diameter growth at given periods on the upper end of each section, the diameter breasthigh, the total height and the clear, used and merchantable lengths. The observations determine the class, form and condition of the tree. Although a tree analysis may include many combinations of the above measurements, two important classes are distinguished: a stump analysis includes measurements of the diameter growth at given periods upon the stump only, no matter what other measurements it may comprise; a section analysis includes measurements of the diameter growth at given periods upon more than one section. When, in a stump or section analyses, the measurement of the diameter growth at given periods covers only a portion of the total diameter growth, the analysis is a partial stump analysis, or a partial section analysis.

Tree class. All trees of approximately the same size. The following tree classes are distinguished: seedling, shoot, small sapling, large sapling, small pole, large pole, standard, veteran.

Tree crown. That part of a tree that is branched, forming a head.

Trench planting. A method of planting on dry ground, in which the seeds of young trees are set in pits or trenches.

Triangular planting. A method of planting in which the unit of arrangement is an equilateral triangle, at each apex of which young trees are placed.

Turkey, n. A bag containing a lumberjack's outfit. To "histe the turkey" is to take one's personal belongings and leave camp.

Two-storied forest. Comprising on the same area two classes, which vary considerably in height, composed of trees of different species. The term is not applicable to forest under reproduction, in which the appearance of two stories is the temporary result of an incomplete process, but to those forests of which the two stories of growth are a natural and permanent feature. In a two-storied forest the taller trees form the overwood, or upper story. The shorter trees form the underwood, or lower story.

Underbrush. All large, woody plants, such as witch-hobble, laurel, striped maple and devil's club, which grow in a forest, but have no main stem or trunk.

Undergrowth. The ground cover, underbrush, and young trees below the large sapling stage.

Undercut, n. The notch cut in a tree to determine the direction in which the tree is to fall, and to prevent splitting.

Underplant, v. To plant trees under an existing stand.

Used length. The sum of the lengths of logs cut from a tree.

Used volume. The sum of the volumes of logs cut from a tree.

Valuation area. A forest area of known size upon which measurements or other detailed studies are made for the determination of the stand or yield.

Valuation survey. The measurement or other detailed study of the stand upon a valuation or experiment area. Two kinds of valuation survey are distinguished: 1. The strip survey comprises the measurement of a stand, or a given portion of it, upon strips usually 1 chain wide. 2. The plot survey comprises the measurement of the stand, or a given portion of it, upon isolated plots not in the form of strips.

Veteran. A tree over 2 feet in diameter breasthigh.

Veteran forest. A forest of veterans.

Volume. Amount or mass of a tree or stand.

Volume table. A tabular statement of the volume of trees in board feet or other units upon the basis of their diameter breasthigh, their diameter breasthigh and height, their age, or their age and height.

Volunteer growth. Young trees which have sprung up in the open, as white pine in old fields, or cherry and aspen in burns.

Wanigan, n. A houseboat used as sleeping quarters or as kitchen and dining room by river drivers.

Wedge a tree, to. To topple over with wedges a tree that is being felled.

Weed. A plant out of place; not of any appreciable economic value.

Weed tree. A tree of a species which has little or no value.

White water man. A log driver who is expert in breaking jams on rapids or falls.

Windbreak. 1. The breaking of trees by wind.
2. A belt of trees, which serves as a protection from wind.

Windfall. 1. A tree thrown by wind.
2. An area on which the trees have been thrown by wind.

Wind-firm. Able to withstand heavy wind.

Work, v. To harvest the final yield under a working plan.

Working, n. The harvesting of the final yield under a working plan. Working is annual when cuttings are made each year; periodic when they are made after uniform periods of two or more years; and intermittent when they are made at irregular intervals. Sustained annual, periodic or intermittent workings are those under which the amount of wood cut is so regulated that the productive capacity of the forest does not decrease but produces a substained yield, which likewise may be annual, periodic, or intermittent.

Working area. The total forest area managed under a working plan.

Working plan. A detailed and comprehensive scheme for the best permanent use of a forest.

Working plan renewal. The preparation of a new working plan for a given tract, when the present working plan has been carried out, or changed conditions require its revision.

Yarding donkey. A donkey engine mounted upon a heavy sled, used in yarding logs by drum and cable.

Yield. The amount of wood at present upon, or which after a given period will be upon, a given area.

INDEX.